Identification of *Marine and Freshwater* Molluscs Shells

The Author

Prof. Dr. A. D. Dholakia is well known expert authority in the field of Fisheries Science related literature. His first book was published in 1986 titled "*Fish Processing Preservation and Quality Control*" in Gujarati, which fetched Sahitya Academy Award from Government of Gujarat, the first of its kind in literature related to fish industry. His second publication was "*Fisheries and Aquatic Resources of India*" in 2004, an English book marketed at International level. His third book on "*Ornamental Fish Culture and Aquarium Management*" in 2009, an English book is being marketed at International level. The book has been given FIVE STAR ranking on internet for standards of book was his fourth book "*Identification of Prawns/ Shrimps of India and their Culture*" in 2010. His fifth book "*Delicious Seafood Recipes*". This is his sixth book.

Prof. Dr. A.D. Dholakia is M. Sc., LL. B. (Sp) and Ph. D. in Marine Aquaculture, with Marine Prawn culture subject. He is having rich 32 years experience in the field of fisheries in different departments. He was positioned in very important posts during his tenure like:

- ☆ Selected by GPSC and posted as Processing Technologist, Head, Shark Liver oil Factory Govt. of Gujarat (GOG) Veraval and Extraction unit of Porbundar
- ☆ Handled ICAR (Govt. of India) Projects on different subjects.
- ☆ Member ISI Standard Fixing Committee of Govt. of India for Fish Products
- ☆ Head Fisheries Training Institute, GOG, Veraval.
- ☆ Senior Manager, Freezing Plant, GFCCA, Veraval.
- ☆ Head Marine Prawn Hatchery, GOG. Okha.
- ☆ Head, Aquaculture Department, College of Fisheries, JAU, Veraval.
- ☆ In charge of Project on Effect of pollution on coastal biomass with Ecological Commission of GOG, Vadodara.
- ☆ Head, Fisheries, (Molluscs) Research Station, JAU, Sikka for more than 7 years.
- ☆ Consultant to GEER Foundation GOG, for Socio Economic Survey Project.
- ☆ His work on pearl oyster culture was appreciated by "Z News T.V," "D.D.-11" and local T.V.
- ☆ His eight episodes are being repeatedly shown in National T. V. programme and more than 16 radio talks from All India Radio have been broadcasted.
- ☆ Examiner for M.Sc., Bhavnagar University, Master's degree for Mahana Pratap Univ. of Agriculture & Technology, Udaipur, (Rajasthan). B.F.Sc. for Junagadh. Agricultural Univ. Junagadh, Gujarat and Rajendra Agricultural Univ. Bihar, Pusa (Samastipur).

The author has published 26 International, National level and 47 State level research papers/articles, which is unmatchable evidence of his expertise in the field. After retirement he is sharing his knowledge, experience and hobbies with students of schools and colleges of Gujarat at free of cost.

Identification of *Marine and Freshwater* Molluscs Shells

Dr. A.D. DHOLAKIA
M.Sc., LL.B. (Sp), Ph. D.
(Retd.) Research Officer and Head
Fisheries Research Station,
Junagadh Agricultural University, Sikka
Gujarat, India

2025
Daya Publishing House®
A Division of
Astral International Pvt. Ltd.
New Delhi – 110 002

First Published, 2013
Reprinted, 2025

ISBN: 978-81-7035-760-5 (Hardbound)

Publisher's note:

Every possible effort has been made to ensure that the information contained in this book is accurate at the time of going to press, and the publisher and author cannot accept responsibility for any errors or omissions, however caused. No responsibility for loss or damage occasioned to any person acting, or refraining from action, as a result of the material in this publication can be accepted by the editor, the publisher or the author. The Publisher is not associated with any product or vendor mentioned in the book. The contents of this work are intended to further general scientific research, understanding and discussion only. Readers should consult with a specialist where appropriate.

Every effort has been made to trace the owners of copyright material used in this book, if any. The author and the publisher will be grateful for any omission brought to their notice for acknowledgement in the future editions of the book.

Published by : **Daya Publishing House®**
A Division of
Astral International Pvt. Ltd.
– ISO 9001:2015 Certified Company –
47636/23, Ansari Road, Darya Ganj
New Delhi - 110 002
Phone: 011-4354 9197, 2327 8134
E-mail: info@astralint.com
Website: www.astralint.com

Laser Typesetting : **Classic Computer Services**
Delhi - 110 035

Digitally Printed at : **Replika Press Pvt. Ltd.**

Acknowledgement

I dedicate this book to my wife Smt Sangita to whom I received utmost encouragement and appreciation. But for the help received from her to keep myself free from any domestic chore whatsoever, the writing of the book would not have been impossible. Also to my mother Late Smt Jayshreeben and Father Shri Dwijendraray who were my inspiration.

I can not forget to mention the name of my student, co-worker, Shri Pinkesh Patel Ex. Sr. Research Assistant presently working in Canada, who worked with me when we were collecting molluscs from sea shore. He helped in identification of molluscs shells from Gujarat sea shore.

I am thankful to Dr. H.B. Dave and Shri V.J. Thakker for encouraging and supporting the work and zeal of providing information for students, research workers and to shell collectors.

Prof. Anshuman D. Dholakia

Foreword

Many authors published on molluscs of different genera. Scientific and detailed study also published on molluscan species. In beginning of 19th century Jems Hornell published book on Economic importance of Molluscs of Okha Mandal (Kathiawar).

Dr A. D. Dholakia made wonderful attempts for making guide book for molluscan shell collectors and naturalist's. This book will be useful to scientist for further research, students and shell collectors as a hand book.

I congratulate Dr. Dholakia giving useful information about marine and freshwater shells. He deserves the gratitude of all connected with molluscan research and shells collectors as a hobby.

Dr. H.B. Dave
Deputy Director of Fisheries
Govt. of Gujarat
Rajkot

Preface

The shell of molluscs, "Chhip" is always attraction of collection when one visits sea shore. Collectors would like to know about the details of such shells, because of varieties of shells. Some enthusiastic collectors would search for different size and shape of such shells.

Very less information on identification of such shells, including classification is available. In this book information of classification, identification, in most case with photographs is given. Anatomy, their system like nerves system, locomotion, Reproduction, Respiratory, etc are also included. The availability of such molluscs is also tabulated. Glossary of terms used in identification of molluscs is given to facilitate identification.

The details of commercial uses of molluscs in the field of food, Medicines, Ayurvedic medicine, trade goods, tools, Architecture, Decorations and religion is given.

These informations along with identification, will help students, research workers and shell collectors. However, I will be happy to receive any suggestion. Please write to me on my residential address.

Prof. Anshuman D. Dholakia

201, Shashwat Apartments,
Opp. Sarthi 3 Bunglows,
Near Vaibhav Laxmi Dham,
Memnagar, Ahmedabad - 380 052

Recommendation

I know Dr. Dholakia since 1976 when we were in Department of Fisheries, Govt. of Gujarat. Once upon a time when I was Head of Jamnagar Office, Gulf of Kutch was under me, I was thinking the need of such work. Identification of shells requires much concentration. I congratulate Dr. Dholakia for writing such a valuable book giving information on molluscs. This will give guidance to students, research workers and shell collectors to identify molluscs with classification. His contribution in Fisheries of Gujarat is very valuable and unforgettable. I wish him every success in his life.

V.J. Thakker
Retd. Deputy Director of Fisheries,
Govt. of Gujarat

Contents

Chapter 1
Introduction

The shell of molluscs, "Chhip" is always attraction of collection when one visits sea shore. Collectors would like to know about the details of such shells, because of varieties of shells. Some enthusiastic collectors would search for different size and shape of such shells.

Shell collectors now and in the past form an important group of naturalist who have made a significant contribution to MALACOLOGY, the study of molluscs. They have been particularly active in the study of taxonomy through their extensive collections and in the mapping of the distributions of different species. All collectors have trend to concentrate on the shell.

Molluscs form the second–largest group in the animal kingdom (after the arthropods which include the insects, crustaceans and spiders), with a great diversity of species. These animals are important economically; some, like oysters, mussels and abalones, form the basis of fisheries in many parts of the world. They have biomedical and veterinary importance. They have been used as currency and for adornments, as musical instruments and as ritual objects. Molluscs are conventionally divided into six classes: Monoplacopora, Amphineura, Gastropoda, Scaphopoda, Bivalvia and Cephalopoda

The *non-marine molluscs of India* are a part of the molluscan fauna of India. There are 5070 species of marine and non-marine molluscs living in the wild in India. There are 3371 species of marine molluscs in India. There are 1671 species of non-marine molluscs living in the wild in India. This includes 1488 terrestrial species in 140 genera and 183 freshwater species in 53 genera.(Alfred, J.R.B. (1998) Faunal Diversity in India: An Overview: In Faunal Diversity in India, i-viii, 1-495. (Editors. Alfred, JRB, *et al.*, 1998). ENVIS Centre, Zoological Survey of India, Calcutta.)

Type of Shell

Mollusc shells are of two main shapes with a multitude of minor variations. Some are coiled, as in the group known as gastropods, and others are secreted in two halves, the bivalves. The latter can be confused with similar looking shells produced by two other groups of animals; the ostracods and the brachiopods. The ostracods, a wider spread group of crustaceans, have a small bivalve shell enclosing their bodies, which bear jointed limbs (their legs are not obvious, particularly in preserved specimens, unless one of the valve is dissected off) The brachiopod's valves are held together by a stalk which protrudes through one of them and their soft parts are totally different from those of a mollusc, including muscles which both open and close the shell.

The bivalves, unlike the branchiopods, have only adductor muscles, which close the valves and at the same time compress an elastic ligament. When the muscles relax the compressed ligament opens the shell. For this reason bivalve shells always gape open when they die.

The Anatomy of Molluscs

In bivalves and gastropods the body lacks any solid skeletal elements. Instead of being attached to bone or ligaments the muscles work against the "fluid skeleton" formed by the blood which thus aids locomotion.

The modern gastropod shell is much more likely to be a very subtle exoskeleton against which the soft-bodied gastropod can extend itself out from its shell by an unscrewing action, using its hydrostatic skeleton and the muscles of the mantle. The coiled shell would therefore be an essential feature in the development of the gastropod, enabling it to withdraw into its shelter when facing a predator, or when environmental conditions are adverse. It also enables the animal to extend itself even when not immersed in water, by providing a series of inclined planes to operate against when extending its body.

Molluscs of all types can usually be identified as processing, either singly or in combination, the following characteristic organs:

Shell

The typical calcareous exoskeleton of molluscs, secreted by the mantle with its shape typical of the class, or subdivision, within the molluscs. They are often coiled as in most gastropods, or formed from two halves as in the bivalves.

Mantle

The extensive epithelium that covers most of the animal and secretes the shell. In gastropods and bivalves it is the tissue nearest the shell, in sea slug and cephalopods it is the tissue that covers most of the back of the animal.

Gut

A coiled tube that starts at the mouth and ends at the anus. Food passed along it is digested by enzymes secreted in the gut tube. Particles and soluble materials are taken into cells by a process known as pinocytosis. All molluscs have a gut and various organs are associated with it, often including paired salivary glands and a large digestive gland, both types of gland secrete digestive enzymes.

Redula

A unique, tongue -like molluscan feeling organ. Used like a flexible rasp it can tear off pieces of vegetable or animal tissue form food making it suitable for injection into the gut. The structure of the redula is typical for different groups of molluscs but it is not found either the bivalves or scaphopods. In most animals it is only just visible to the naked eye with the dimensions of a thread of cotton 0.25 X 20 mm. The redula enable animals such as chitons and limpets to feed on algae that encrust rocks and shores, or winkles to feed on seaweeds. Modifications to the basic pattern have enabled the whelks to become scavengers and predators and, with the cone shells. In the most bizarre modification of all, the teeth can be loaded with poisons and used to kill such fast moving animals as fish, on which some species feed exclusively.

Gills

Feather-like structures found in the mantle cavity of most molluscs that are used primarily for oxygenating the blood. Clean and well oxygenated water is circulated through the mantle cavity by the beating action of cilia on the gills which keep the water moving across their surfaces. The gill surfaces absorb oxygen from the water and pass carbon dioxide out into it. Gills are used for oxygenation of the blood in all groups of molluscs except the scaphopods, which do not have gills, while the bivalves also use them to sieve their food from the water current drawn into the shell by the gill cilia.

Siphon

A tube formed by a specialization of the mantle margin, found in the carnivorous gastropods, such as the whelks, and in many bivalves. In gastropods water is drawn into the mantle cavity through the siphon and exhaled through a hole in the mantle, where as in bivalves there are usually two siphons, an inhalant and an exhalant siphon. Siphons in many bivalves can be very large and when extended be several times longer than the shell, the length being related to the depth at which the animal normally lives in the sand. Animals with short siphons lives in shallow burrows and have only a small "pallial sinus" where deep forms like many true clams have enormous siphons and very deep pallial sinuses

Byssus Threads

Anchor thread spun by the foot from a secretion from the byssus gland and only found in bivalves. Some bivalves live cemented to rocks for the whole of their adult lives and others are free-living in the sand; lay the bivalves that produce byssus threads. They are well developed in certain species, such as the mussel *Mttilus* sp. and Ear shell, *Pinna* sp

The animal can dissolve this thread which has the remarkable property of being able to harden under seawater. Such glue could be very useful to man, especially perhaps for dentists, who could use it to glue filling into teeth. (Peter & Ptricia Newell Seashells, 1979, p. 13 – 14)

Seashells have acted as money; they serve as ornaments in our homes and adorn our bodies as necklaces, cameos and other jewellery. They have been used in magic and as musical instruments, as drinking cups and in the making of fine porcelains.

Shell Collection, Cleaning and Preservation

Shells are often found encrusted with calcareous algae, sea mats, barnacles, calcareous worm tubes and other organisms which may add to the interest of shell but more often obscure the shell colours. Careful scraping is the best way to remove them but is only practicable with species having a fairly smooth and regular surface.

Decomposing mollusc tissues have a particularly unpleasant smell and shells should be carefully cleaned, inside as well outside before they are brought into your house. In some part of the world, near Cape Town for instance, small ants will efficiently remove the soft tissues, but careful work with a jet of hot water and a piece of wire is usually the most effective method.

Many people treat shells as household ornaments and make very attractive displays with them, but this is to run the risk of sunlight bleaching out the colours. For the serious shell collector is probably better to use a system of drawers with shells packed in individual boxes with ported on cotton wool. Such a system is costly, but when starting a collection a little planning may enable a collector to build up a flexible storage system which could be expanded with the collection. Perhaps a system of drawers manufactured for anor purpose could be purchased, for example, the type of drawers used by engineers to store different sorts of screws and nuts. A neat label giving the site of collection, date and the animal's full name is essential.

If chitons are to be displayed in a collection then they require special treatment. One of the best ways is to remove the soft parts, as far as possible, by scraping them away with small scalpel or knife and then tying the remaining of the animal to a thin piece of board with thread. The flattened animals can then be immersed in a solution of 50 per cent alcohol or 10 per cent formalin in sea water for about two weeks and then removed and allowed to dry. They can then be cut free from the board and the specimen is ready for labelling and putting away in the appropriate part of your collection. (Peter and Patricia Newell Seashells p. 41-42)

The type of shell found in each class is typical and members of each class can be identified from the accompanying simple key shown below.

Key to the Most Common Classes of Molluscs

Sl.No.	*Details*	*Class*
1.	Shell coiled and in one piece	*Gastropoda*
2.	Shell coiled and in piece, chambered and vary pearly on the inside	*Cephalopoda*
3.	Shell a tapering and curved white tube	*Scaphopoda*
4.	Shell composed of eight small plates	*Amphineura*
5.	Shell composed of 2 halves, often of roughly similar size, but mirror images of each other, often with a hinge and ligament	*Bivalvia*

Contd...

Contd...

Sl.No.	Details	Class
6.	Shell a shallow cone.	*Gastropoda*
7.	Internal shells horny often reduced	*Gastropoda or Cephalopoda*
8.	Internal shells an elongated oval plate composed of numerous fine chambers.	*Cephalopoda*
9.	Shell tubular. If molluscs then probably a wood boring bivalve, otherwise this may be formed by a worm.	*Bivalvia*

Source: Peter and Patricia Newell Seashells, p. 42.

Cephalopods

This group is mostly fast swimming predators with well developed brains and sense organs. Many have internal shells but the pearly nautilus *Nautilus pomilus,* has a true external one. Cephalopods are identified by their body shape and they all have a number of head tentacles armed with suckers.

Two types of shell shape, both evolved in the Cambrian period, characterize modern cephalopods; an internal shell, as in cuttlefish and the external shell as in the nautilus. Both types offer little protection against predators, nor do they enable the animals to survive periods of environmental hostility, but they do dorm a rigid platform for the attachment of the powerful muscles of the mantle which pump water out of the mantle cavity through a nozzle (the funnel), which drives the animal through the water. Their shell also acts as highly efficient and sophisticated flotation devices.

The first cephalopods are known from later Cambrian formations. They were small about 25 mm (1 inch) long and secreted narrow, straighter or slightly curved, chambered shells. The size of the chambers and the weight of the shell probably meant that these shells were not very buoyant. Many grew much larger reaching 40 mm, 70 mm, and some up to 220 mm respectively.

Cephalopods like squid, cuttlefish and octopus are far more mobile and active as compared to other molluscs. Squids are reported to range from less than one inch to 70 feet or more in length when arms are extended. They occur from the surface of sea water to great depths. Cuttlefish mainly inhabit shallow coastal waters and they are found in most tropical and temperate seas. Generally Octopus are found in shallow coastal waters but some species also occur at great depths. Maharashtra, Kerala, Tamil Nadu and Andhra are the main maritime states that contribute higher percentages of cephalopods. Gujarat is next to them.

In *Cephalopoda* class, Squid (*Loligo* spp), in Loliginidae family; Cuttle fish (*Sepiella, Sepis, Sepia* spp.), in Sepiidae family; and *Octopus* spp. In octopodidae family are included. However, as in this book, main concentration is given to shell; details of these species are not included.

The book begins with an account of how molluscs sculpt and pattern their mobile homes in calcium. Classification of molluscs, their important groups like bivalves, cephalopods gastropods, and scaphopods etc are also covered. The main concentration of this book is to provide Identification of shells and molluscs found on Gujarat coast. Details of availability of these molluscs are also discussed.

MARINE MOLLUSCS SHELLS

Chapter 2
Pelecypoda (Bivalves)

Introduction

One of the important classes in molluscs is Bivalvia (Pelecypoda). The body lacks any solid skeletal elements. Instead of being attached to bone or ligaments the muscles work against the 'fluid skeleton'. In this chapter how to identify different families of Bivalves are explained with shape of shell, anatomy, locomotion system, reparatory system, feeding etc. Classification, taxonomy details of their body etc are also given in this chapter.

Key to Families

Remarks on key characters: features used in this key only apply to species included in this guide; they do not consider a few exceptions within the families, the inclusion of which would make the key too complex for general use.

1a. Only 1 adductor muscle scar in each valve .... → **2**

1b. Two (sometimes 3) adductor muscle scars in each valve .... → **6**

2a. Interior of shell partly nacreous, with a non-nacreous border developed at least ventrally. .... **Figure A**

2b. Interior of shell, if nacreous, without a non-nacreous border .... → **3**

3a. Dorsal margin drawn out into ear-shaped or wing-shaped lateral ansions. .... **Figure B**

3b. Dorsal margin not drawn out into such expansions .... → **4**

4a. Ligament mainly internal .... **Figure C**

4b. Ligament only external .... **→ 5**

5a. Hinge with teeth .... **Figure D**

5b. Hinge without teeth .... **Figure E**

6a. Shell with calcareous accessory plates or tube and pallets; a finger-like apophysis projecting from the umbonal cavity in each valve .... **Figure F**

6b. Shell without calcareous accessory plates or tube and pallets; no finger-like apophysis projecting in either valve .... **→ 7**

7a. Anterior and posterior adductor scars very unequal, the anterior one always small .... **Figure G**

7b. Anterior and posterior adductor scars not very unequal .... **→ 8**

8a. Hinge with numerous alternating small teeth and sockets, all or part of them transverse to dorsal margin .... **Figure H**

8b. Hinge not as above .... **→ 9**

9a. Internal ligament present .... **→ 10**

9b. Internal ligament absent .... **→ 11**

10a. Hinge without teeth .... **Figure I**

10b. Hinge with well-developed teeth .... **Figure J**

11a. Hinge teeth and corresponding sockets more or less parallel to dorsal margin .... **Figure K**

11b. Hinge teeth and corresponding sockets not parallel to dorsal margin .... **→ 12**

12a. Shell more than twice longer than high, widely gaping anteriorly and posteriorly .... **Figure L**

12b. Shell not as above .... **→ 13**

13a. Anterior adductor scar elongate, with an oblique ventral lobe detached from pallial line .... **Figure M**

13b. Anterior adductor scar, if elongate, not with an oblique ventral lobe detached from pallial line .... → **14**

14a. Hinge with 3 cardinal teeth, at least in the left valve .... **Figure N**

14b. Hinge with no more than 2 cardinal teeth in either valve .... → **15**

128 Bivalves

15a. Pallial line with a sinus .... **Figure O**

15b. Pallial line without a sinus .... → **16**

16a. Main sculpture of radial ribs .... **Figure P**

16b. Main sculpture concentric; radial ribs, when present, crossed by stronger concentric ridges .... **Figure Q**

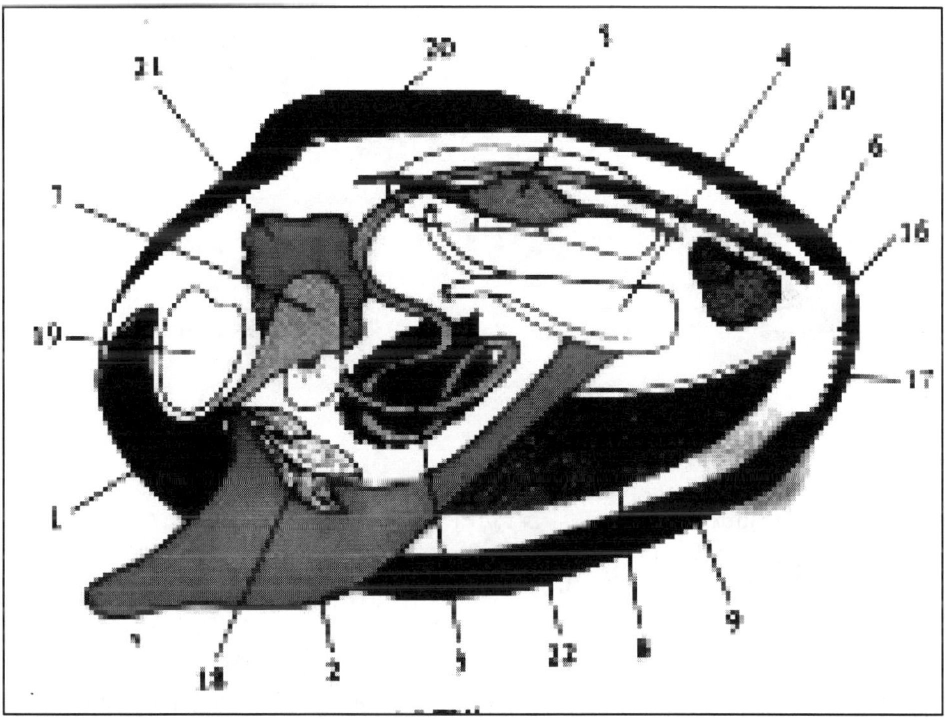

Anatomy of Bivalve

1: Mouth; 2: Foot; 3: Gonad; 4: Kidney; 5: Heart; 6: Anus; 7: Gut; 8: Gill; 9: Mantle; 10: Radula; 11: Operculum 12: Head tentacles; 13: Feeding rentacles; 14: Retractor muscle; 15: Inhalant; 16: Exhalant siphon; 17: Inhalant muscle; 18: Palps; 19: Adductor muscle; 20: Hinge; 21: Digestive gland; 22: Shell.

These animals often have a heavy, hinged calcareous shell which may be composed of identical halves, as in the mussel, *Mytilus edulis* or dissimilar halves, as in scallop or oyster. Because they are sedentary or slow - moving, they have an entirely different internal anatomy from that of gastropods. They do not have a well-defined head and in most species neither the central nervous system nor the sense organs are well developed, although in some species, including the scallop and the cockle, eyes are found on the mantle margins.

Key to Families (Details with Figures)

Figure A

Isognomonidae

Shell compressed, with a straight dorsal margin, slightly inequivalve. Ligament external, set in a series of transverse grooves along the dorsal margin. Hinge without teeth. Interior partly nacreous. Only 1 adductor muscle scar, with a well-developed pedal retractor scar. Pallial line without a sinus.

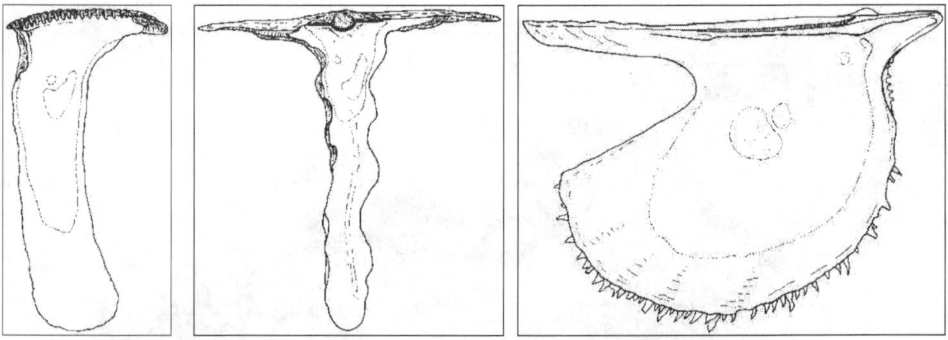

Isognomonidae **Malleidae** **Pteriidae**

Figure A

Malleidae

Shell compressed, irregular in form, more or less elongate dorsoventrally. Dorsal margin often produced at both ends into long, wing-like ears. Ligament set on a transverse median groove. Hinge without teeth. Interior partly nacreous. Only 1 adductor muscle scar, usually with a well-developed pedal retractor scar. Pallial line without a sinus.

Pteriidae

Dorsal margin often produced at each end into a wing-like ear, sometimes very long behind. Shell slightly inequivalve. Right valve with a byssal notch anteriorly. Hinge toothless or with denticles. Interior brilliantly nacreous. Only one adductor muscle scar. Pallial line without a sinus.

Figure B

Limidae

Shell equivalve, higher than long, inequilateral, extended obliquely in an anteroventral direction. Dorsal margin with 2 small ears. Trigonal cardinal area with a median ligamental groove. Hinge toothless. A single adductor muscle scar. Pallial line without a sinus.

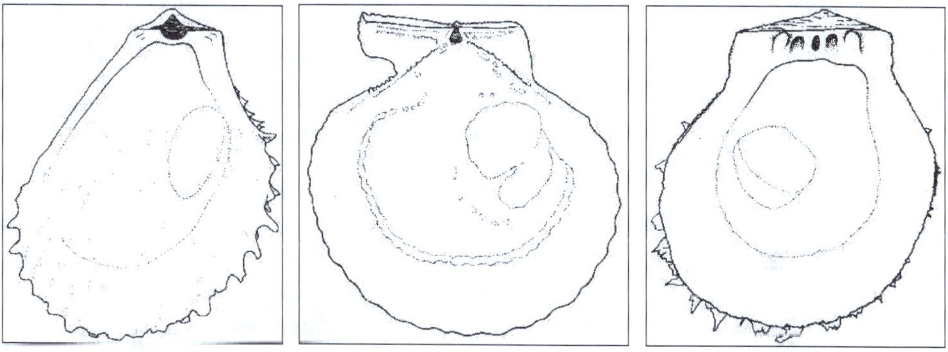

Limidae **Pectinidae** **Spondylidae**

Figure B

Pectinidae

Shell more or less inequivalve, ovate to subcircular with a straight dorsal margin forming wing-like ears. A byssal notch and a ctenolium at right valve. Ligament internal, in a small trigonal pit pointing under the umbones. Hinge without teeth. A single adductor muscle scar. Pallial line without a sinus.

Spondylidae

Shell stout, usually inequivalve and cemented to substrate by the right valve. Hinge line straight. A trigonal cardinal area, higher in the right valve than in the left. Ligament internal. Hinge with 2 strong teeth and 2 deep sockets in each valve, symmetrically arranged in relation to the internal ligament. A single adductor muscle scar. Pallial line without a sinus.

Figure C

*Anomiidae

Shell inequivalve, often irregular, adhering to substrate by means of a calcified byssus passing through a hole-like embayment of right valve. Ligament internal. Hinge without teeth. Central area of the interior thickened, with 1 or 2 retractor muscle scars in left valve, in addition to the single adductor scar. No pallial sinus.

Placunidae

Shell thin, rounded to saddle-shaped, very compressed laterally, slightly inequivalve. Ligament internal, forming an inverted V-shaped structure. Hinge without teeth. A single adductor muscle scar. Pallial line without a sinus.

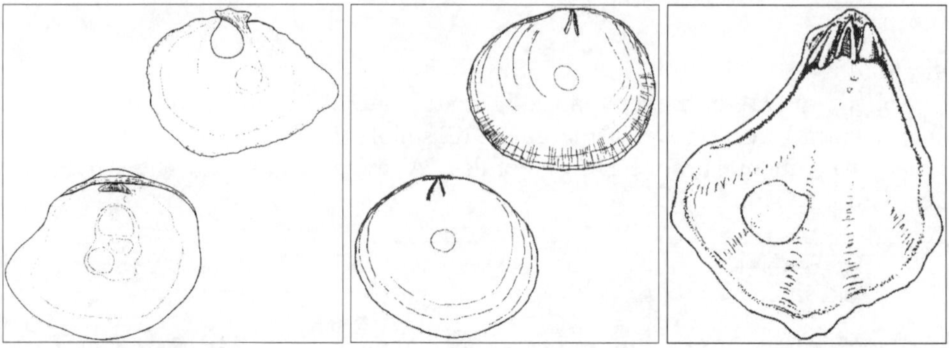

Anomiidae **Placunidae** **Plicatulidae**

Figure C

*Plicatulidae

Shell slightly inequivalve, cemented to substrate by the right valve. Cardinal area small. Ligament internal. Hinge with 2 crenulated teeth and 2 sockets in each valve, symmetrically arranged in relation to the internal ligament. A single adductor muscle scar. No pallial sinus.

Figure D

Tridacnidae

Shell equivalve, thick, heavy and often very large, with strongly scalloped free margins. Umbones ventral, free margins of the valves dorsal-most in position. Byssal gape, when developed, internally plicate. Outer surface with strong radial folds. Ligament external. Hinge with ridge-like cardinal and lateral teeth. A single adductor muscle scar, associated with a pedal retractor scar, submedian in position. Pallial line without a sinus.

Figure E

Gryphaeidae

Shell more or less inequivalve, cemented to substrate by the left valve, with a microscopic vesicular structure. Ligamental area with a shallow median groove.

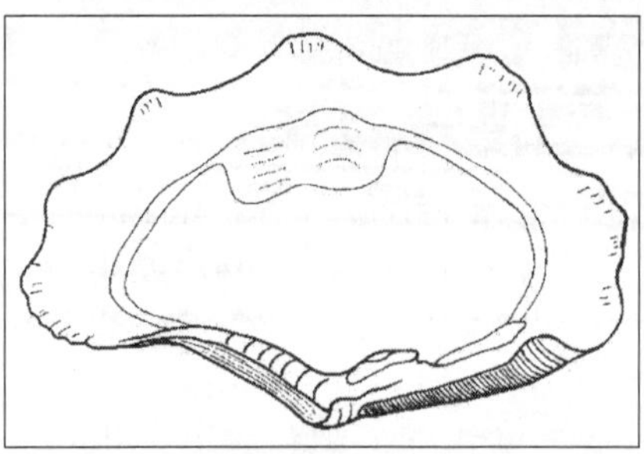

Tridacnidae

Figure D

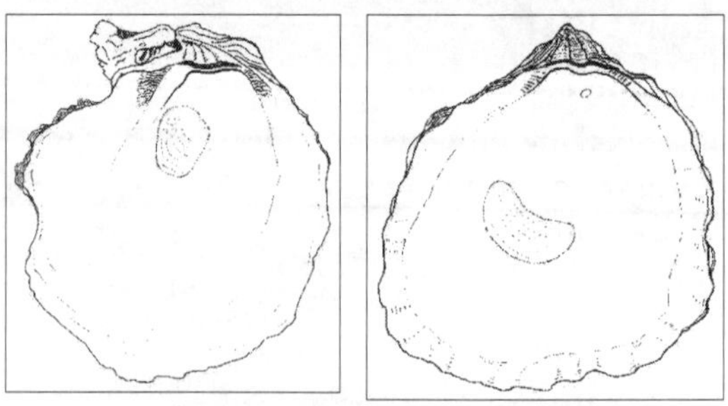

Gryphaeidae **Ostreidae**

Figure E

Hinge without teeth. A single adductor muscle scar, closer to the hinge. Internal margins with long, branched, sinuous chomata.

Ostreidae

Shell inequivalve, cemented to substrate by the left valve, right valve quite flat. Ligamental area with a shallow median groove and 2 lateral thickenings. Hinge without teeth. A single adductor muscle scar, median in position or nearer to the ventral margin. Internal margins smooth or with simple short chomata.

Figure F

Pholadidae

Shell subequivalve, gaping. Dorsal margin forming an umbonal reflection. A number of accessory calcareous plates about the main shell. Ligament reduced. Hinge without teeth. A finger-like internal apophysis is present. Three adductor muscle scars. Pallial line deeply sinuated.

Teredinidae

Shell reduced, equivalve, widely gaping. Anteroventral margin with a deep, right-angled notch. Dorsal margin forming an umbonal reflection. Ligament reduced. Hinge without teeth.

A finger- like internal apophysis. An internal umbonoventral ridge, with a knob at both ends. Three adductor muscle scars. Accessory calcareous tube lining burrow long, closed by a pair of pallets.

Figure G

Mytilidae

Shell equivalve and very inequilateral, with a byssal gape. Umbones at or near anterior end. Periostracum prominent. Ligament external, deep-set, supported by a whitish ridge. Hinge teeth absent or reduced. Adductor muscle scars unequal, the anterior one small to absent. Pallial line without a sinus. Inner side with an extensive nacreous layer.

Pinnidae

Shell brittle, equivalve, subtrigonal, ventrally and posteriorly gaping; very inequilateral, pointed in front. Anterior end eroded and internally closed by small transverse partitions. Ligament linear. Hinge without teeth. Interior with a thin nacreous layer, restricted to the anterior half. Two unequal adductor muscle scars.

Figure H

Arcidae

Shell equivalve or slightly inequivalve, mostly longer than high, more or less inequilateral. Umbones prosogyrate, on top of a wide cardinal area. Ligament external, often with V-shaped grooves. Hinge elongate, almost straight, with numerous small transverse teeth. Two subequal adductor muscle scars. Pallial line without a sinus.

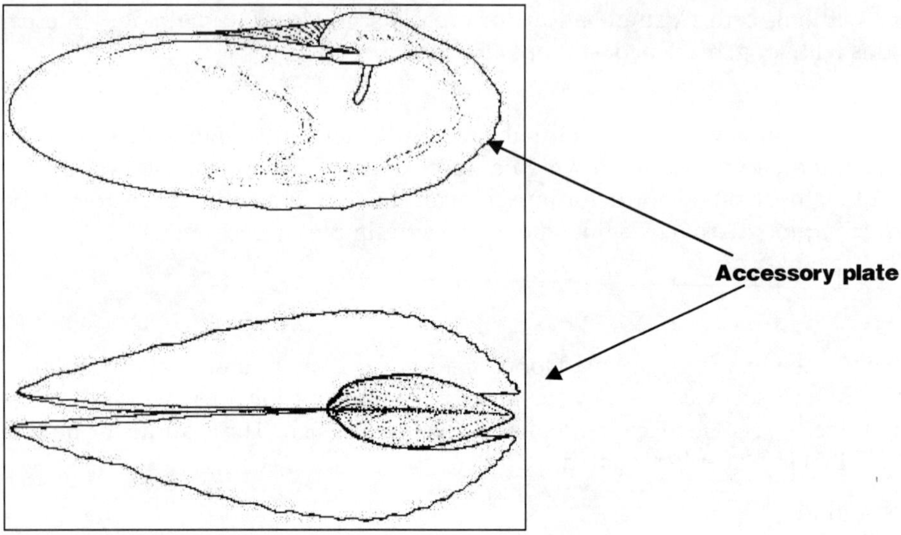

Pholadidae

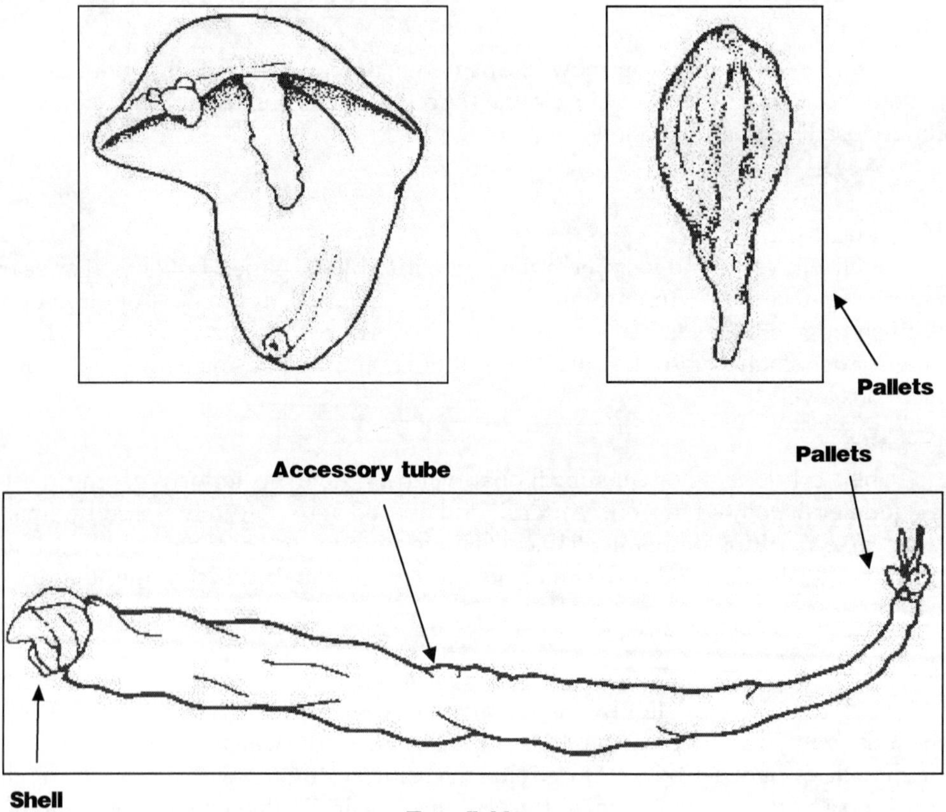

Teredinidae

Figure F

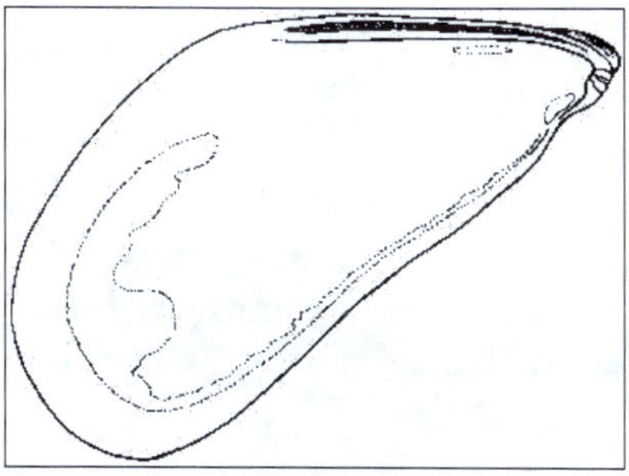

Mytilidae

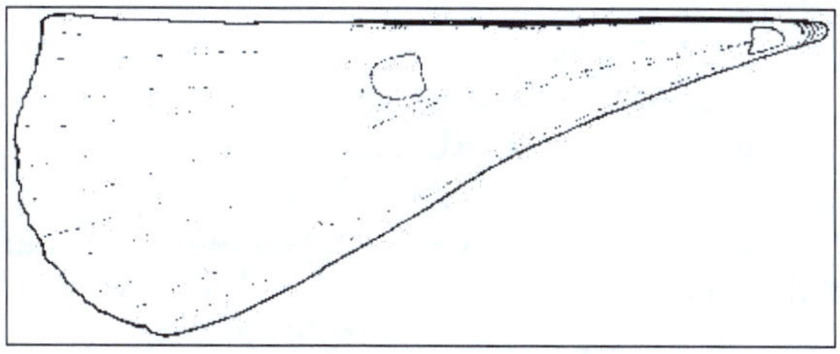

Pinnidae

Figure G

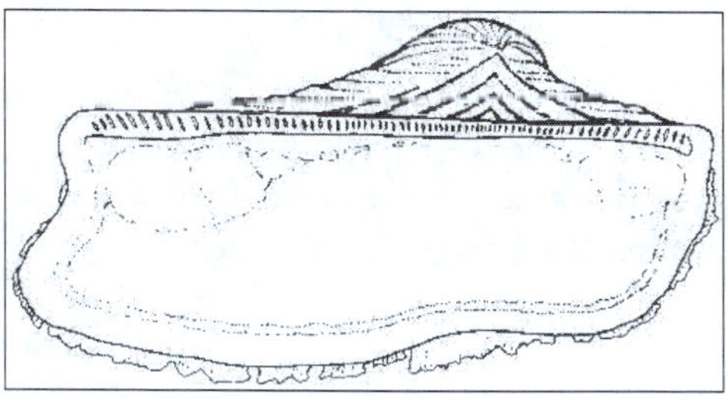

Arcidae

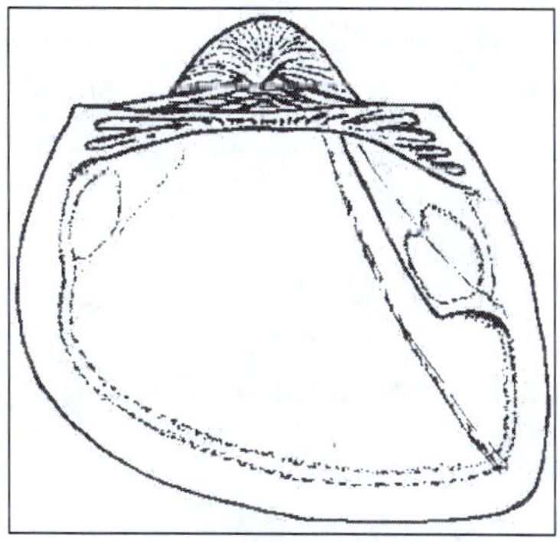

Cucullaeidae

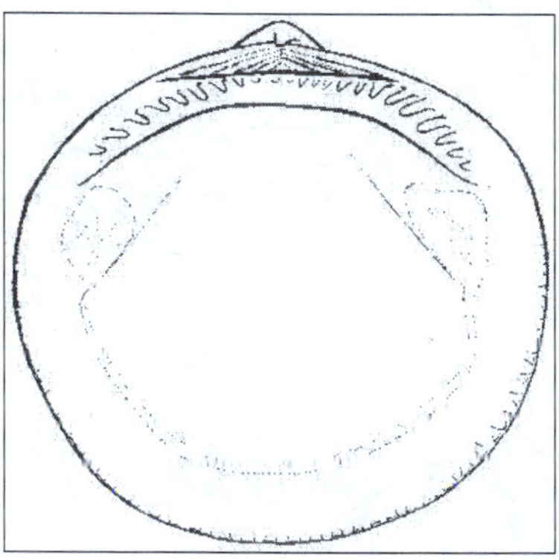

Glycymerididae

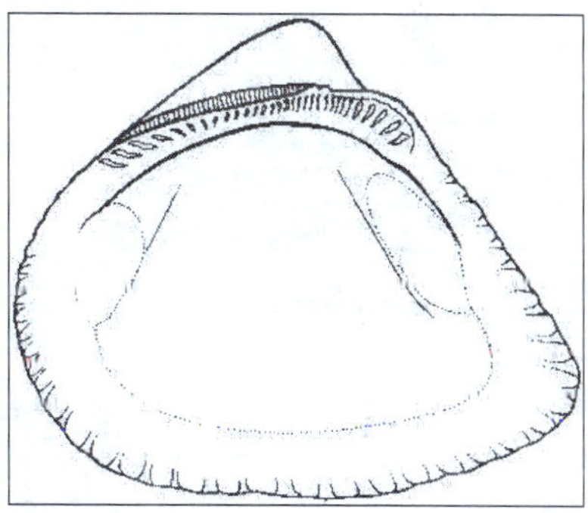

Noetiidae

Figure H

Cucullaeidae

Shell inflated, inequilateral, slightly inequivalve. Umbones subcentral, on top of a trigonal cardinal area with chevron-shaped grooves and external ligament. Hinge elongate, straightish, with a series of transverse teeth, and subhorizontal teeth at both ends. Two subequal adductor muscle scars, inner margin of posterior scar on a projecting shelf. Pallial line without a sinus.

Glycymerididae

Shell equivalve, closed, subequilateral, rounded in outline. Submedian umbones, on top of a trigonal cardinal area engraved by tent-shaped grooves and covered with external ligament. Hinge arched, bearing a series of teeth diverging outwards. Two subequal adductor muscle scars, their inner margin with a radial ridge. Pallial line without a sinus.

Noetiidae

Shell equivalve, generally inequilateral and longer than high. Umbones often opisthogyrate, set apart by a trigonal cardinal area. Ligament external, with oblique grooves and transverse striations. Hinge elongate, straightish, with numerous small transverse teeth. Two subequal adductor muscle scars with a ridge or a shelf along 1 or both scars. Pallial line without a sinus.

Figure I

Laternulidae

Shell thin and brittle, elongate-ovate, truncate to rostrate posteriorly, gaping, subequivalve. Umbones with an obvious median slit. Outer surface finely granulated. Internal ligament attached on protruding spoon-like pits, each supported by an oblique

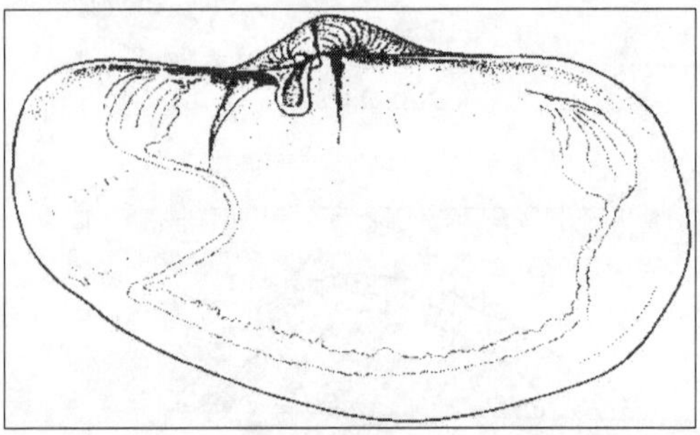

Laternulidae

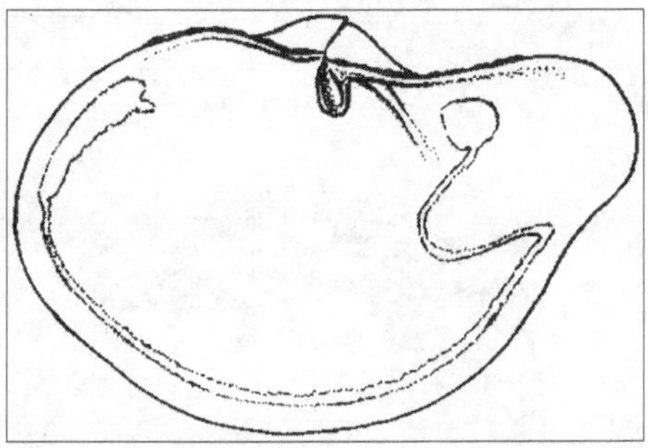

Peneplomatidae

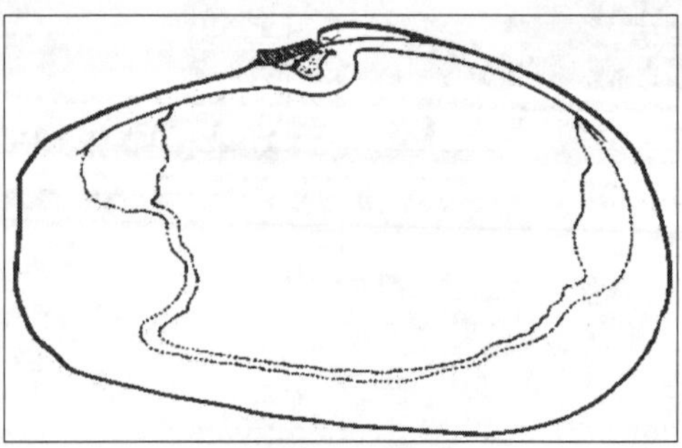

Thracugae

Figure I

radial ridge. Hinge without teeth. Interior subnacreous. Two adductor muscle scars. Pallial line with a broad sinus.

*Periplomatidae

Shell rounded, subrostrate posteriorly, gaping, markedly inequivalve. Umbones with an obvious median slit. Outer surface finely granulated. Internal ligament attached on protruding spoon-like pits, each supported by an oblique buttress. Hinge without teeth. Interior subnacreous. Two adductor muscle scars. Pallial line with a rather deep sinus.

*Thraciidae

Shell thin, elongate-ovate, truncate posteriorly, usually closed, subequivalve. Umbones without an obvious median slit. Outer surface finely granulated. Internal ligament attached on trigonal pits, not protruding ventrally nor supported by oblique buttresses. Hinge without teeth. Interior not nacreous. Two adductor muscle scars. Pallial line with a broad sinus.

Figure J

*Crassatellidae

Shell thick, equivalve. Umbones prosogyrate to orthogyrate. Sculpture mainly concentric. Lunule and escutcheon distinct. Internal ligament in a pit of hinge plate. Hinge with 2 cardinal teeth and lateral teeth. Two adductor muscle scars. Pallial line without a sinus.

Mactridae

Shell equivalve. Umbones prosogyrate. Internal ligament well developed, in a trigonal pit of hinge plate. Hinge characteristic, with 2 cardinal teeth and lateral teeth; cardinal teeth of the left valve forming an inverted V-shaped process. Two adductor muscle scars. Pallial line with a well-developed sinus.

Mesodesmatidae

Shell equivalve, inequilateral, subtrigonal to wedge-shaped. Umbones opisthogyrate. Internal ligament in a deep pit of hinge plate. One or 2 cardinal teeth and lateral teeth. Two adductor muscle scars. Pallial line with a short sinus.

*Semelidae

Shell rather compressed, often slightly inequivalve, with a rightwards flexure posteriorly. Internal ligament in a small pit of hinge plate. Hinge with 2 cardinal teeth and lateral teeth. Two adductor muscle scars. Pallial line with a deep sinus. Cruciform muscles leaving small paired scars near pallial line.

Figure K

Chamidae

Shell thick, strongly inequivalve and inequilateral; very variable in shape, cemented to substrate by either the left or the right valve. Umbones prosogyrate, low, spirally wound. Ligament external. Hinge with large, curved teeth and corresponding sockets, more or less parallel to dorsal margin. Two subequal adductor muscle scars. Pallial line without a sinus.

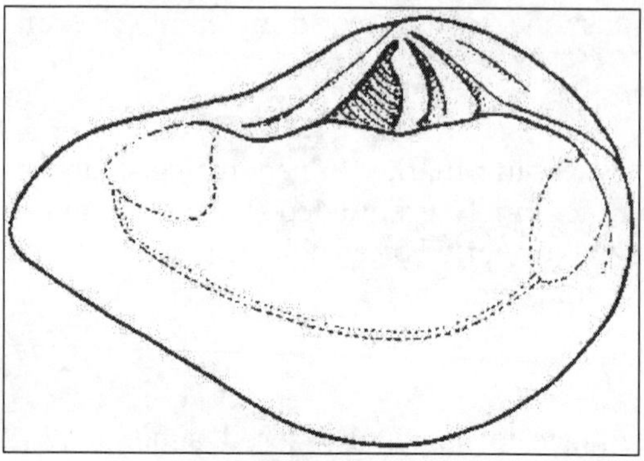

***Crassatellidae**

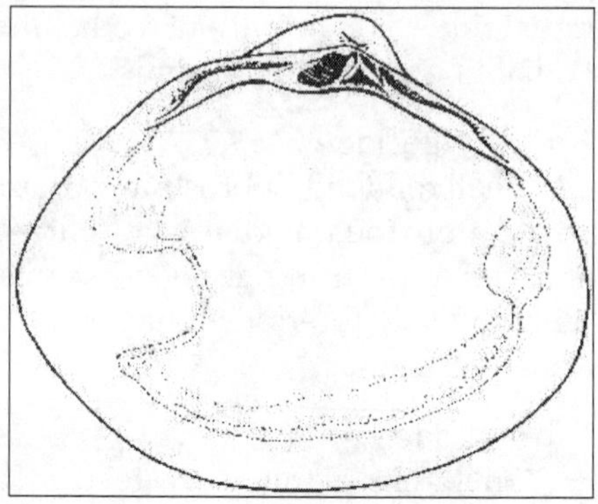

Mactridae

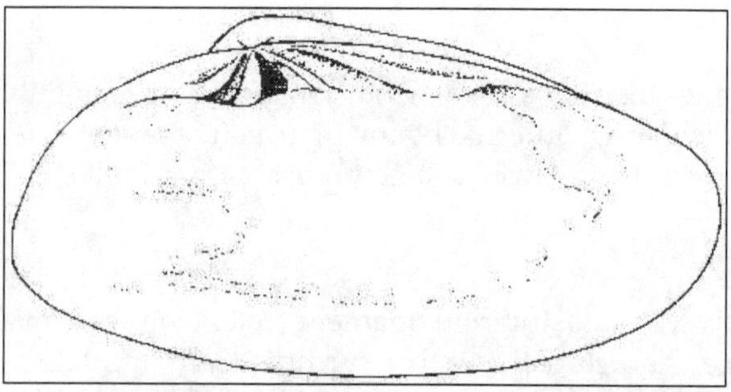

Mesodesmatidae

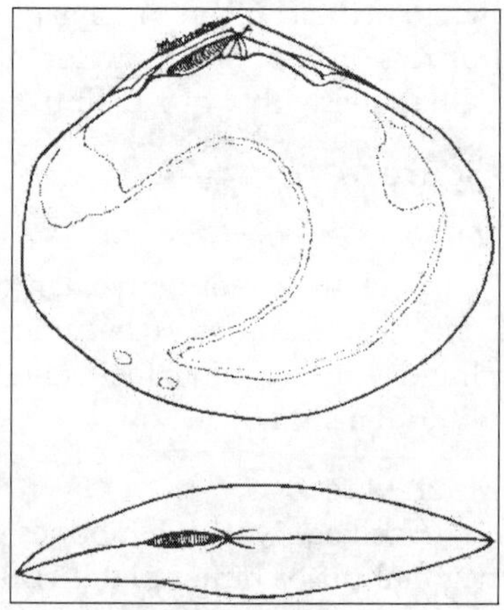

Semelidae

Figure J

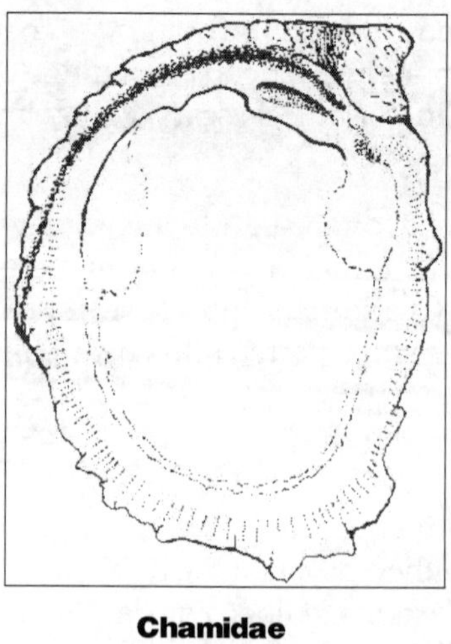

Chamidae

Figure K

Figure L

Solecurtidae

Shell equivalve, elongate-quadrate, widely gaping at both ends. Umbones subcentral.Ligament external, on projecting nymphs. Two cardinal teeth in either valve. Two adductor muscle scars. Pallial sinus deep.

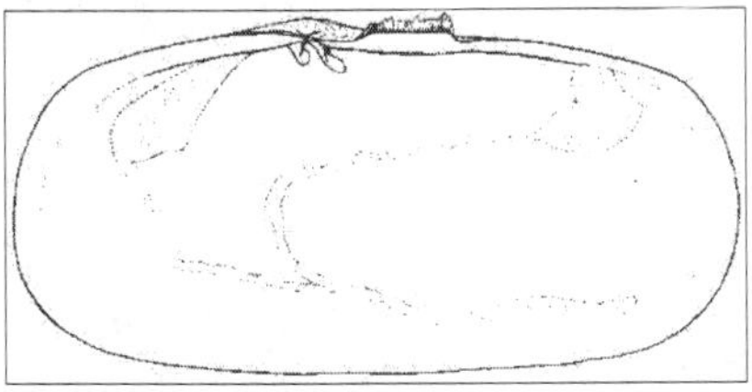

Solecurtidae

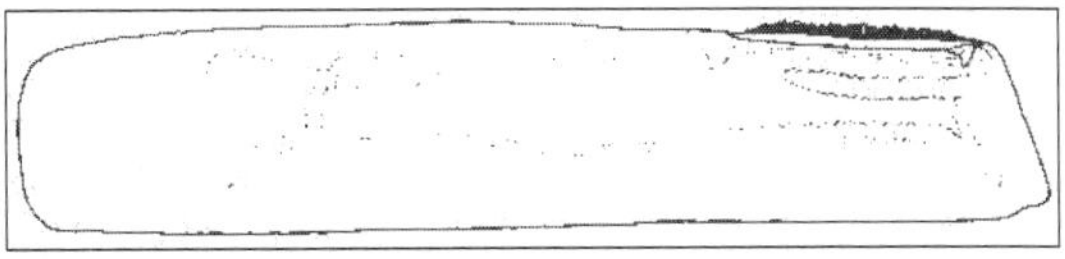

Solenidae

Figure L

Solenidae

Shell equivalve, with a narrowly elongate shape, gaping at both ends. Umbones more or less near the anterior end. Ligament external. Hinge feeble. Two adductor muscle scars, the anterior one larger. Pallial sinus relatively shallow.

Figure M

Lucinidae

Shell equivalve, lenticular, slightly inequilateral. Ligament more or less deeply sunken in posterodorsal margin. Two cardinal teeth and lateral teeth in either valve,

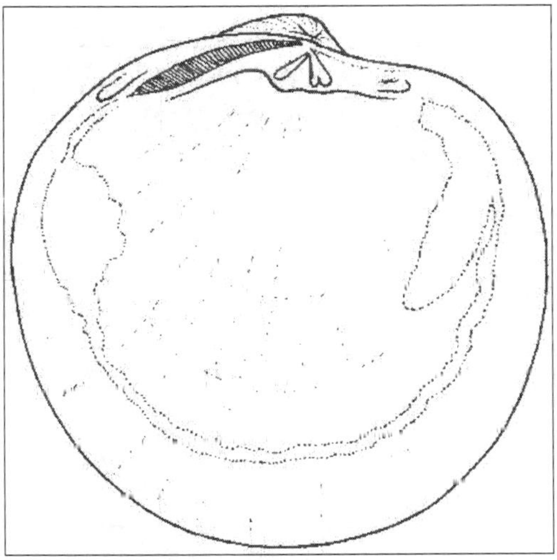

Lucinidae

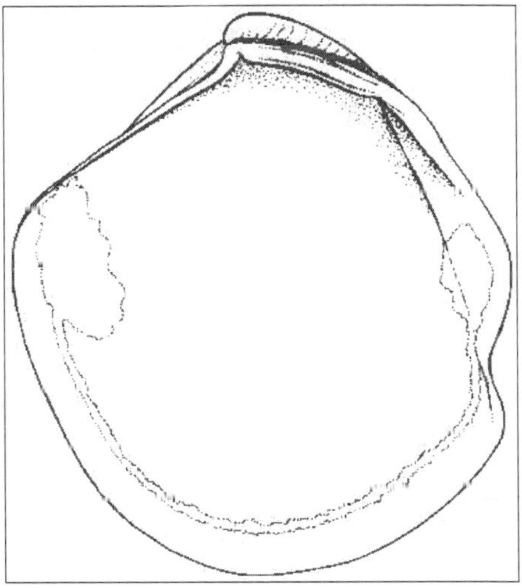

***Thyasiridae**

Figure M

sometimes reduced to absent. Two adductor muscle scars, the anterior narrowly elongate with an oblique ventral lobe detached from pallial line. No pallial sinus.

*Thyasiridae

Shell equivalve, thin, trigonal, inequilateral. Posterior part of valves set off by 1 or more deep radial furrows or folds. Ligament marginal. Hinge teeth obsolete to absent. Two adductor muscle scars, the anterior elongate, with an oblique ventral lobe detached from pallial line. No pallial sinus.

Figure N

Corbiculidae

Shell equivalve, solid, umbones prosogyrate. No lunule nor escutcheon. Periostracum conspicuous. Ligament external. Three diverging cardinal teeth in each valve, and strong anterior and posterior lateral teeth. Two adductor muscle scars. Pallial sinus reduced to absent.

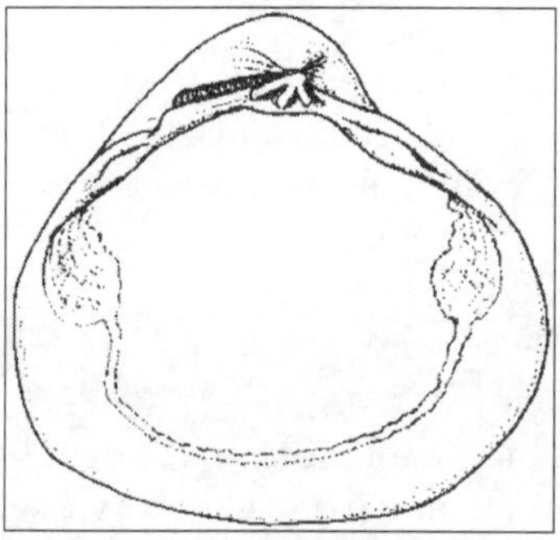

Corbiculidae

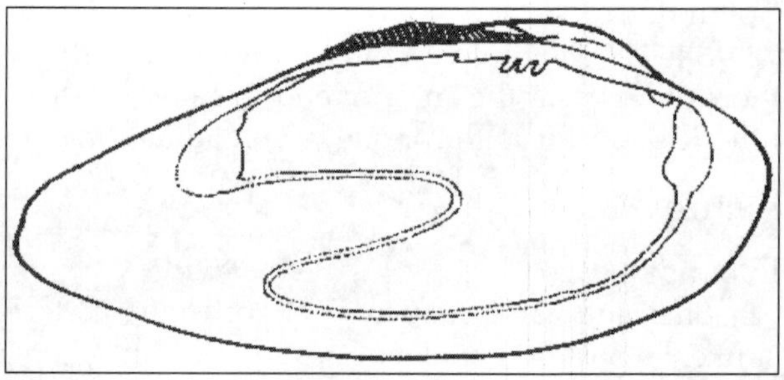

Glauconomidae

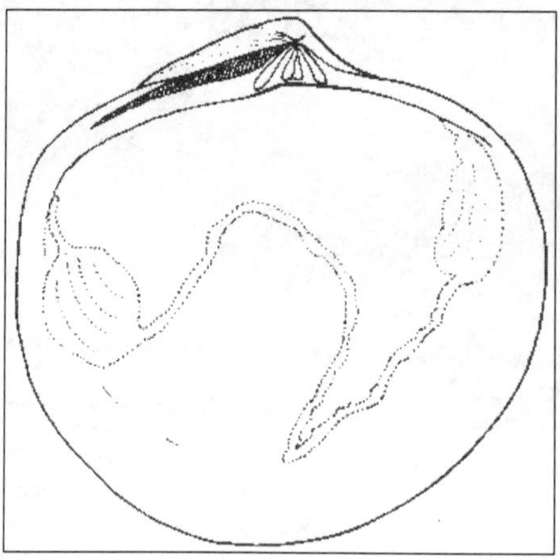

***Petricolidae**

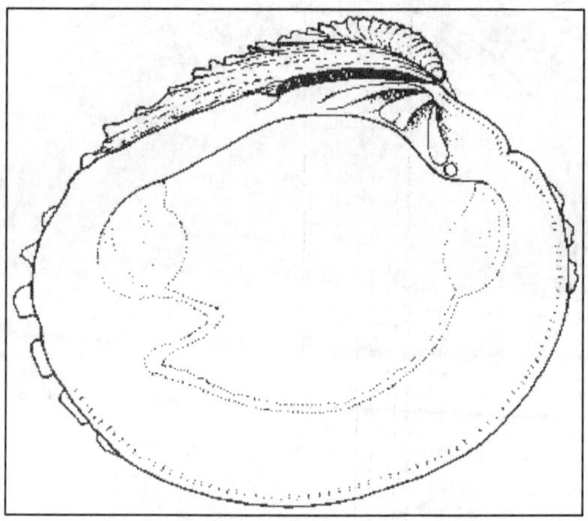

Veneridae

Figure N

Glauconomidae

Shell equivalve, gaping posteriorly, inequilateral. No lunule nor escutcheon. A conspicuous, greenish periostracum. Ligament external. Three cardinal teeth in each valve, lateral teeth wanting. Two adductor muscle scars. Pallial sinus deep and narrow.

*Petricolidae

Shell equivalve, inequilateral, with prosogyrate umbones. No lunule nor escutcheon. Three cardinal teeth in left valve and only 2 in right valve; lateral teeth wanting. Two adductor muscle scars. Pallial sinus deep.

Veneridae

Shell mostly solid, equivalve, inequilateral, with prosogyrate umbones. Lunule and/or escutcheon usually present. Ligament external. Three cardinal teeth in each valve, anterior lateral teeth sometimes present. Two adductor muscle scars. Pallial sinus usually present.

Figure O

Donacidae

Shell equivalve, trigonal to wedge-shaped, with a shorter posterior end. Umbones opisthogyrate. Ligament external. Two small cardinal teeth and lateral teeth. Two adductor muscle scars. Pallial sinus deep. Cruciform muscle scars obscure.

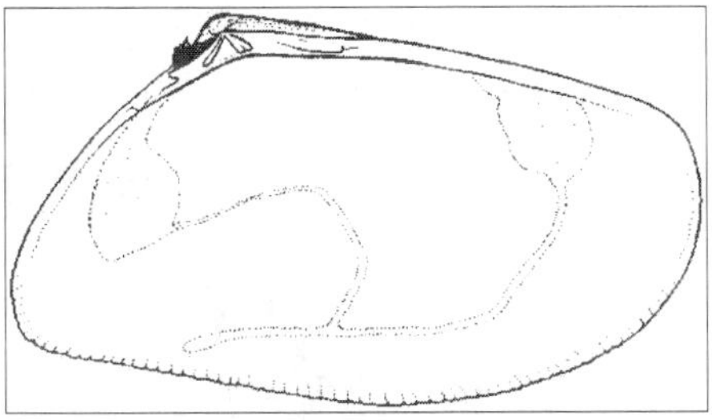

Donacidae

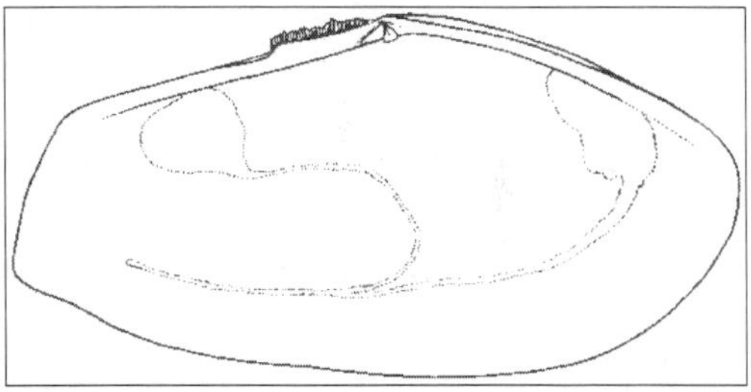

Psammobiidae

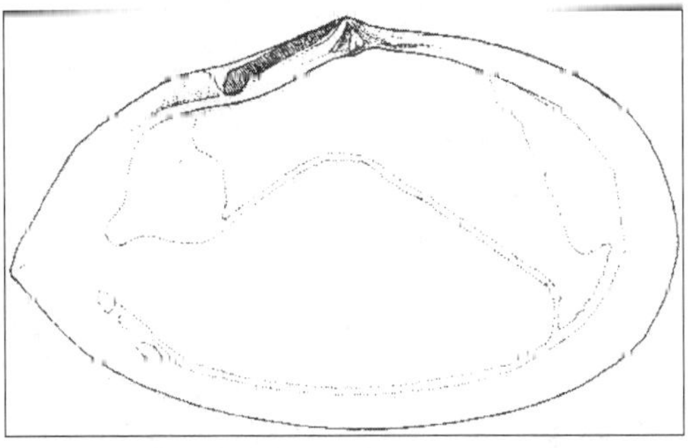

Tellinidae

Figure O

Psammobiidae

Shell ovate to subelliptical or trapezoidal, somewhat gaping. Ligament external, on projecting nymphs. Two small cardinal teeth in either valve; lateral teeth absent. Two adductor muscle scars. Pallial sinus deep. Cruciform muscle scars often obscure.

Tellinidae

Shell rather thin and compressed, often slightly inequivalve, with a rightwards flexure on posterior end. Ligament external. Two small cardinal teeth in either valve; lateral teeth often present. Two adductor muscle scars. Pallial sinus deep. Cruciform muscles leaving small paired scars near pallial line.

Figure P

Cardiidae

Shell equivalve, inflated, oval to subquadrate, sometimes heart-shaped. Umbones prominent.External sculpture mostly radial. Ligament external. Hinge characteristic,

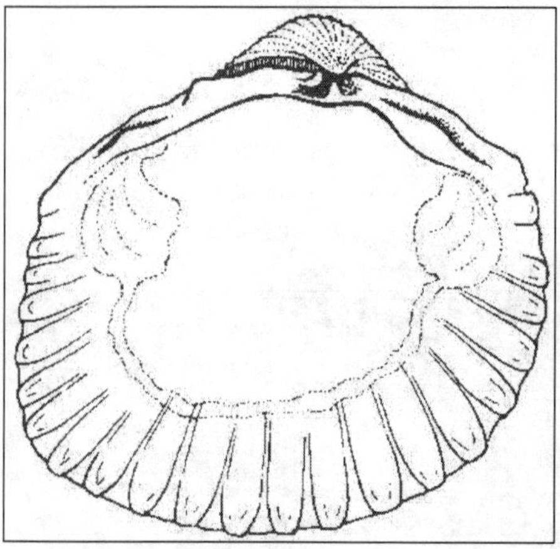

Cardiidae

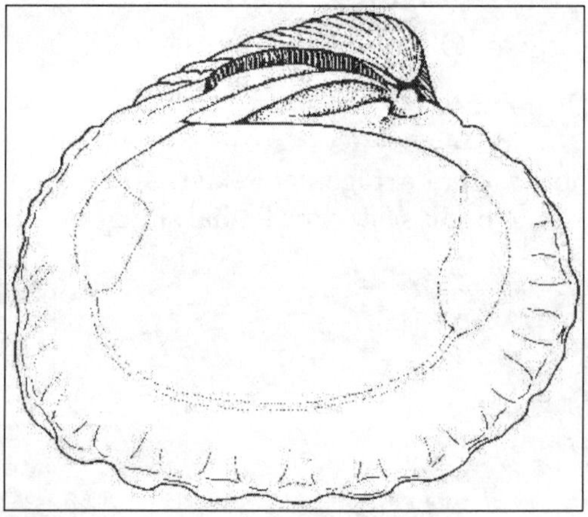

Carditidae

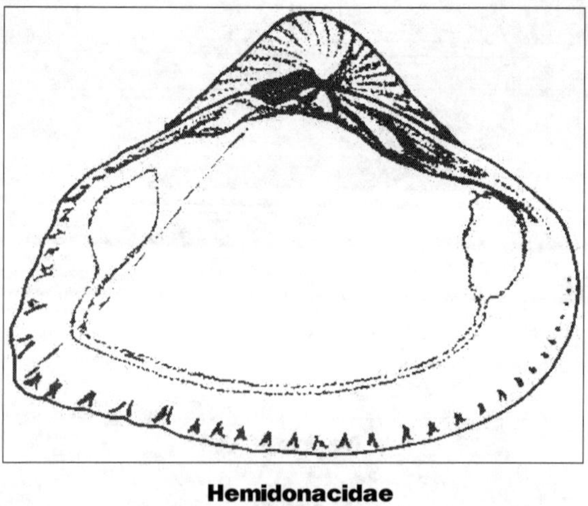

Hemidonacidae

Figure P

with teeth curving outwards; 2 cardinal teeth and lateral teeth in each valve; cardinal teeth cruciform in arrangement. Two adductor muscle scars. Pallial line without a sinus.

Carditidae

Shell equivalve, stout and inflated, inequilateral. Exterior mostly with radial ribs. Ligament external. Two cardinal teeth, unequal and with fine transverse striations; lateral teeth frequently reduced to absent. Two adductor muscle scars. Pallial line without a sinus.

Hemidonacidae

Shell equivalve, subtrigonal to wedge-shaped and transversely elongate, more or less inequilateral. Outer surface with smooth radial ribs, often reduced on the anterior part of shell. Ligament external. Two cardinal teeth and elongate lateral teeth in each valve. Two adductor muscle scars. Pallial line without a sinus.

Figure Q

Fimbriidae

Shell equivalve, inflated, thick, transversely elliptical. Lunule lanceolate, escutcheon narrow. Outer surface with latticed sculpture, concentric ribs more prominent. Ligament marginal. Two massive cardinal teeth, one nearby anterior lateral tooth, and one long, remote posterior lateral tooth in each valve.Two adductor muscle scars. Pallial line devoid of sinus.

*Ungulinidae

Shell equivalve, lenticular, slightly inequilateral. Ligament external, not sunken in a marginal groove. Two cardinal teeth in either valve, lateral teeth reduced to absent. Two adductor muscle scars, the anterior elongate but without an oblique ventral lobe detached from pallial line. No pallial sinus.

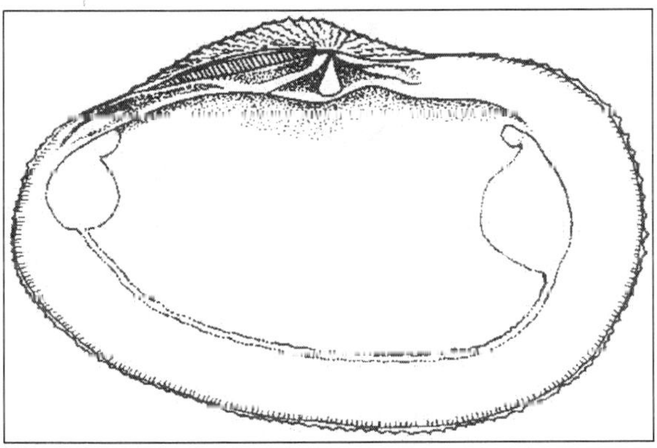

Fimbriidae

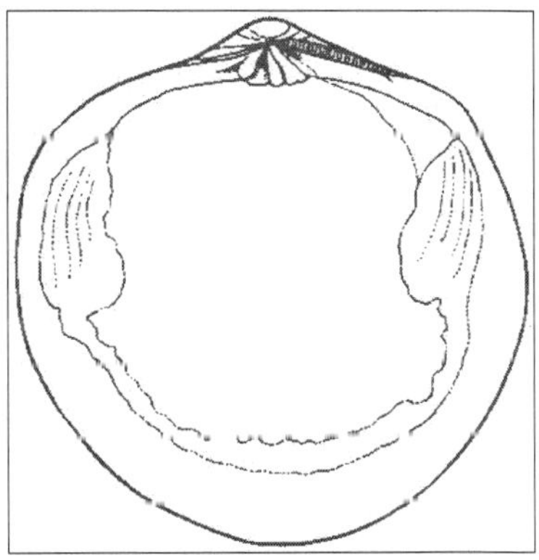

***Ungulinidae**

Figure Q

General Remarks

Bivalves are aquatic molluscs that show a fundamental bilateral symmetry. Their characteristic shell is composed of 2 calcified valves lying on the right and left sides of the body. Both valves are typically equally convex (equivalve shell), but they may differ in size and shape (inequivalve shell) as a result of an alteration of bilateral symmetry.Valves are articulated along a marginal process of the dorsal side called the hinge, and are connected by an elastic and poorly calcified structure, the ligament. Under the action of the ligament, the 2 valves tend to open along their anterior, posterior and mainly ventral margins. They are closed by the pulling action of 1 or 2 (sometimes 3) adductor muscles. These are fixed to the inner side of valves by areas leaving well-defined imprints, the adductor muscle scars.

In the typical morphology of bivalves, some of the general features are:

Shell

The bivalve shell acts as a skeleton to protect against predators, and in burrowing species it helps to keep mud and sand out of the mantle cavity. Its main component is calcium carbonate and is formed by the deposition of crystals of this salt in an organic matrix of the protein, conchiolin. Calcium for shell growth is obtained from the diet, or taken up from seawater. The colour, shape and markings on the shell vary considerably between the different groups of bivalves (Gosling, 2003).

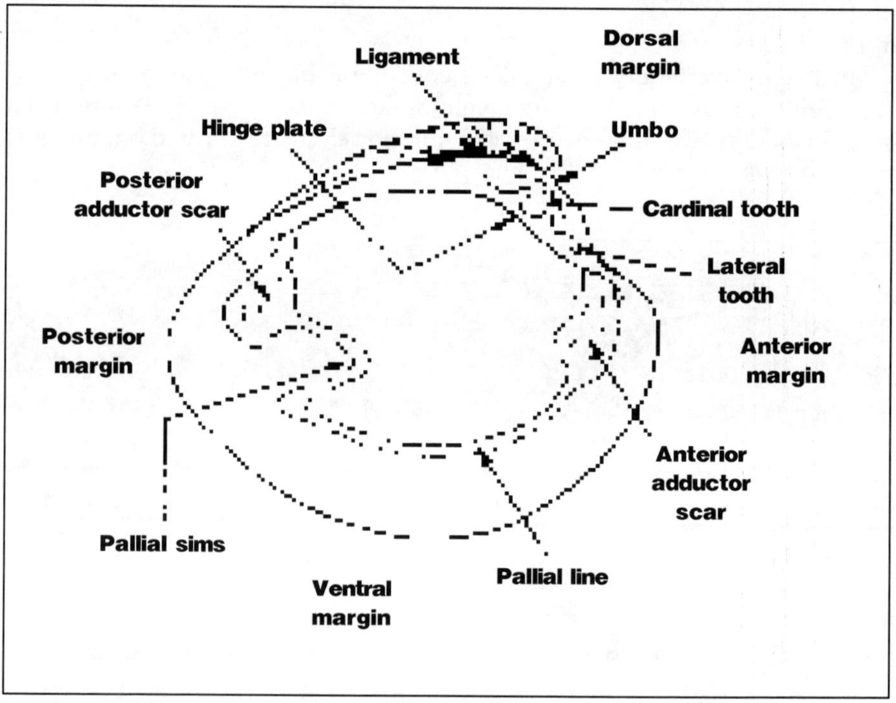

Figure 1: Interior of Left Valve

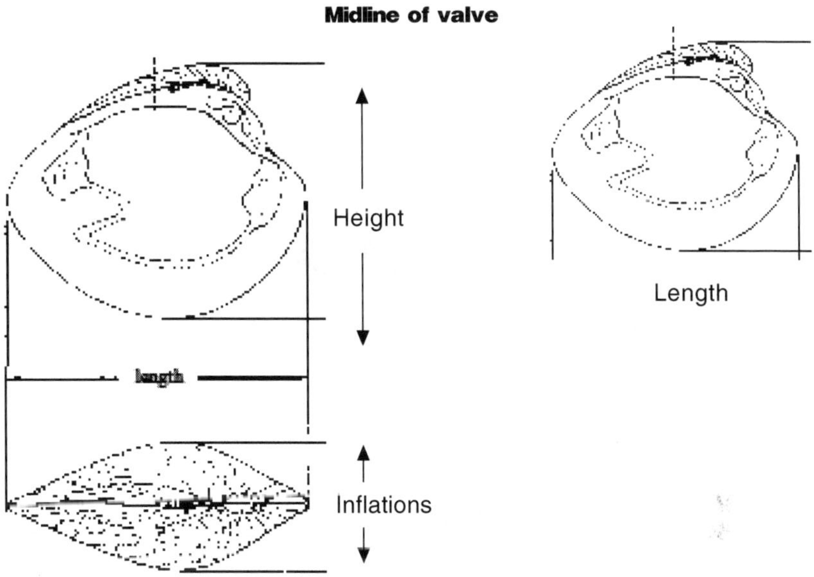

Figure 2

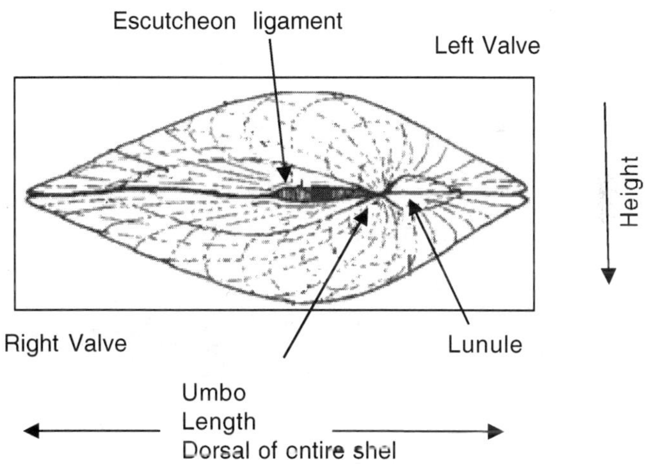

Figure 3

Shell Formation

The shell is secreted by the mantle. The calcium ions excreted from the blood mix with the fluid present in the mantle cavity, forming calcium carbonate. The calcium carbonate is absorbed by "conchiolin", a secretion of the mantle. The conchiolin crystallises into various forms, of which calcite and argonite forms are utilised in shell formation (Apte, 1998).

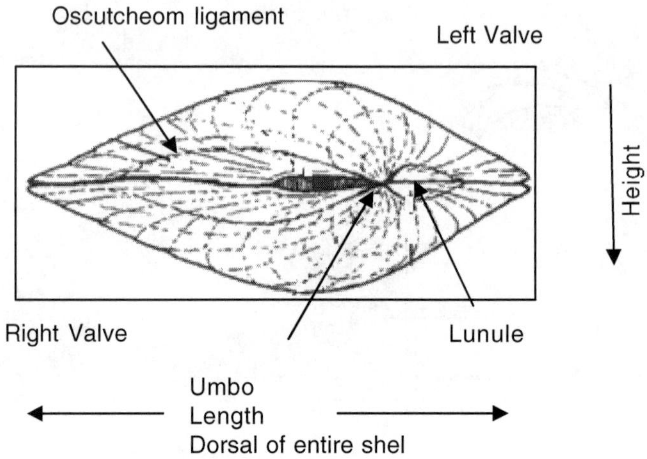

Figure 4

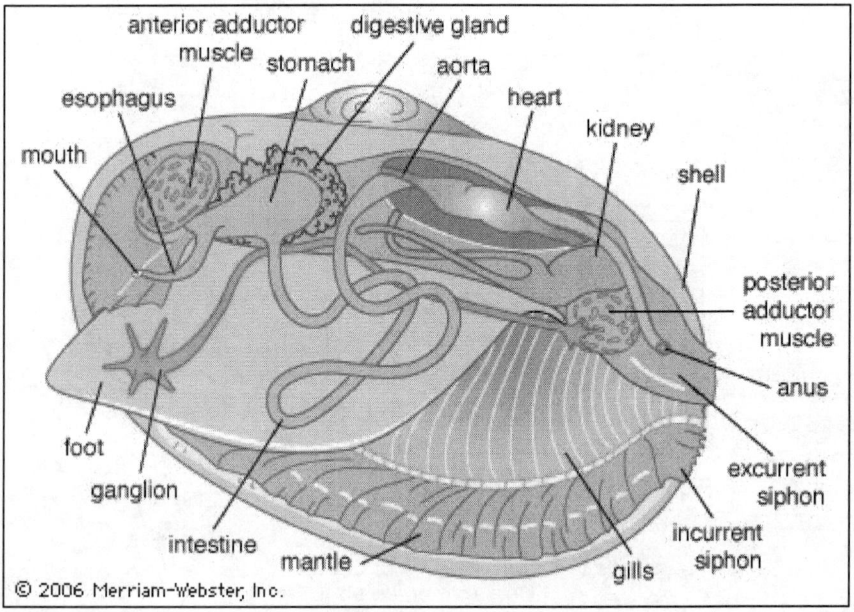

Morphology of Bivalves

Mantle

In bivalves the mantle consists of two lobes of tissue which completely enclose the animal within the shell. Between the mantle and the internal organs is a capacious mantle cavity. Cilia on the inner surface of the mantle play an important role in directing particles onto the gills and in deflecting heavier material along rejection tracts towards the inhalant opening (Gosling, 2003).

Gills

The lamellibranch gills, or ctenidia, are two large, curtain-like structures that are suspended from the ctenidial axis that is fused along the dorsal margin of the mantle. Generally, the gills follow the curvature of the shell margin with the maximum possible surface exposed to the inhalant water flow. Cilia on the gill filaments have specific arrangements and functions. They are responsible for drawing water into the mantle cavity and passing it through the gill filaments, and then upwards to the exhalant chamber and onwards to the exhalant opening. In bivalves the gills have a respiratory as well as a feeding role. Their large surface area and rich haemolymph supply make them well suited for gas exchange (Gosling, 2003).

Stomach

The mouth is ciliated and leads into a narrow ciliated oesophagus. Ciliary movement helps to propel material towards the stomach. Indeed, this method of moving material is found throughout the length of the alimentary canal, primarily because it lacks a muscular wall. The stomach is large and ovalshaped and lies completely embedded in the digestive gland, which opens into it by several ducts (Gosling, 2003).The digestive gland, which is brown or black and consists of blind-ending tubules that connect to the stomach by several ciliated ducts, is the major site of intracellular digestion. Within these ducts there is a continuous two-way flow: materials enter the gland for intracellular digestion and absorption and wastes leave en route to the stomach and intestine. Rejected particles from the stomach as well as waste material from the digestive gland pass into the long coiled intestine. The waste is formed into faecal pellets that are voided through the anus and are swept away through the exhalant opening (Gosling, 2003).

Foot

The primitive mollusc had a broad ciliated flat foot, well supplied with mucous gland cells, and the animal is believed to have moved over the lubricated substrate in a gliding motion, using a combination of ciliary action and muscular contractions. In the evolution of bivalves the body became laterally compressed. Consequently, the foot lost its flat creeping sole and became blade-like and directed in an anterior direction as an adaptation for burrowing. Bivalves use the foot for locomotion and burrow in to substrate (Gosling, 2003).

Gonads

The reproductive system in bivalves is exceedingly simple. The gonads are paired and each gonad is little more than a system of branching tubules, and gametes are budded off the epithelial lining of these tubules. The tubules unite to form ducts that lead into larger ducts and eventually terminate in a short gonoduct (ibid). In primitive bivalves, *e.g.* the nut shell, *Nucula*, the gonoducts open into the kidneys, and eggs and sperm exit through the kidney opening (nephridiopore) into the mantle cavity. In most bivalves the gonoducts open through independent pores into the mantle cavity, close to the nephridiopore. With the exception of oysters (*Ostrea* sp.), fertilisation is external and the gametes are shed through the exhalant opening (Gosling, 2003).

Heart

The heart lies in the mid-dorsal region of the body, close to the hinge line of the shell. It lies in a space called the pericardium, which surrounds the heart dorsally and a portion of the intestine ventrally. The heart consists of a single, muscular ventricle and two thin-walled auricles. The circulatory system is an open system with the haemolymph in the sinuses bathing the tissues directly. From the sinuses the haemolymph is carried to the kidneys for purification (Gosling, 2003).

Excretory Organs

There are two types of excretory organs in bivalves, the pericardial glands and the paired kidneys (in *Mytilus* U-shaped). The brown-coloured pericardial glands, sometimes referred to as Keber's organs, develop from the epithelial lining of the pericardium and come to lie over the auricular walls of the heart. Waste accumulates in certain cells of the pericardial glands and this is periodically discharged into the pericardial cavity and from there it is eliminated via the kidneys (Gosling, 2003).

Nerves

The nervous system of bivalves is fundamentally simple. It is bilaterally symmetrical and consists of three pairs of ganglia and several pairs of nerves. The cerebral ganglia innervate the palps, anterior adductor muscle, and part of the mantle, as well as the statotocysts and osphradia. The pedal ganglia control the foot. The visceral ganglia control a large area: gills, heart, pericardium, kidney, digestive tract, gonad, posterior adductor muscle, part or the entire mantle, siphons and pallial sense organs (Gosling, 2003).

Different Systems

Feed and Feeding

Food is normally planktonic organisms which are filtered out of seawater by mucus-covered gills. The plankton is then removed from the gill by the action of countless cilia which convey it to the mouth as a plankton-laden mucus rope.

Bivalves do not possess a redula. They feed on particles of either plankton or broken-down material suspended in the seawater. The food is separated from the water by a sieving process that takes place on the sticky gills from which food-laden ropes of mucus are conveyed to the mouth. Not all the trapped particles are eaten, some being rejected by a sorting process that occurs on the labial palps, the ciliated flaps that fringe the mouth. Food that is passed into the gut is digested by enzymes produced by the digestive gland and undigested material voided through the anus, which opens in the exhalant siphon. The gills are often enormously developed and are found on both sides of the body between the central foot and the mantle.

Locomotive System

The foot is a producible organ used by burrowing species for digging into the sand. In sedentary forms this is often reduced in size.

Reproductive System

Reproduction in bivalves normally involves the shedding of gametes into the seawater, adults usually being sessile.

The gonads are usually large and the animals are often hermaphrodite. The gamers are shed directly into the water and so mating does not occur.

Nerves System

The nervous system is simple and the head is completely absent, reflecting the sedentary habit. In primitive bivalves (*e.g.*, Palaeotaxodonta) there are four pairs of ganglia–cerebral, pleural, pedal, and visceral. In all other bivalves the cerebral and pleural ganglia are fused into two cerebropleural ganglia, located above and on either side of the esophagus. The pedal ganglia are in the base of the foot, and the visceral ganglia are located under the posterior adductor muscle. Nerve fibres arising from the cerebropleural ganglia extend to the pedal and visceral ganglia.

There is well developed heart and blood vascular system as well as kidney. The mantle margins in deep-burrowing species are extended to form enormous muscular siphons which can extend many times the length of the shell. The shell itself is usually held together by two adductor muscles. One of these has been lost in scallops and so they have only one large adductor muscle, which is placed centrally.

Poisonous Bivalves

In the primitive bivalves the paired gills are small and located posteriorly. The gills in all other bivalves (save septibranches, which have lost their gills) are greatly enlarged and possess a huge surface area. While the gills are thought to serve a respiratory function, respiratory demands are low in these mostly inactive animals, and, since the body and mantle are both bathed in water, respiration probably takes place across these surfaces as well. Such a mechanism has been demonstrated for a few bivalves, most notably freshwater species that are exposed to occasional drought.

Most of the bivalves are edible; the only real risk in eating those concerns the environment from which they have harvested. As far as the author is aware no bivalve is of itself, poisonous to man. This does mean that they not poisonous, and in fact many poisoning occur each year, some of the fatal, after bivalves have been eaten. Without exception bivalves become poisonous only by feeding on, or growing in, water containing toxic material. This can occur in one of two ways: either from material suspended in the wear or from the bioaccumulation of materials dissolved in the water. Both types of accumulation are important as they have a considerable impact on the commercial exploitation of various shellfish stocks.

Bivalve Resources of Marine Creeks of Gujarat

The coastal areas of Gujarat are characterized by numerous parallel streams from the rivers following into the Arabian Sea. The river Bhadar, Ozat, Minsar and Sambi forms deltaic zones as the coastal areas are low lying and gets flooded during the monsoon. During high tide the sea water rushes into these areas through small creeks. River water flow is present only for one month during the monsoon; consequently it creates permanent saline condition creek. Thus there is no estuary as such in the region, but numerous small creeks are found all along the coast. It was observed that wherever there is hard substratum for attachment, some oysters are found, and in some areas of the creek with sandy bottom, clams are observed. They form a good source of food to fisher population, besides serving as an additional income for them in coastal areas of Gujarat, especially during the off fishing season.

Except in the Gulf of Kutch area (Okha, Sikka, Dwarka, Poshitra, Salaya etc) Miani and Navibandar there is no commercial exploitation of bivalve. The other fishing activity area and creeks like Jaleshwar, Chorward, Sutrapada, Vanakbara, Diu, Porbander, bivalve resources were very less or negligible.

Navibander Creek

This creek is about four hundred meters long and hundred meters broad with one opening to the sea and the other end delimited by the check dam to prevent influx of sea water to the interiors. The tidal fluctuation is about 1.5 m and most of the oyster beds are exposed during the low tide. The salinity in the creek ranged from 21 ppt upstream to 34 ppt towards the bar mouth and pH was about 8.1. The soil is clayey in nature a pH range from 6.9 to 7.2.

The oyster resources observed include *Crassostrea gryphoide* (62 per cent), *C. rivularis* (20 per cent), *C. madrasensis* (4 per cent) and *Saccostrea cuculata* (14 per cent). The number of oyster per square meter was 20 – 22 towards upstream and larger numbers of dead ones were also observed in this area. Towards the bar mouth, about 140 – 145 oysters per square meter was observed (Sujatha Thomasm 2005)

Chorward Creek

Chorward creek is about 700 m length, with a maximum depth of 2.5 m. It is about 150 m from the sea. This creek does not have opening to the sea. Generally the soil is sandy except in a few pockets where the soil is silty with high organic load. The salinity ranged from 39 to 40 ppt.

About 15 – 20 oysters per square meter were observed. Majority of the oysters were represented by *C. gryphoides* (70 per cent), followed by *C. rivularis* (20 per cent), *C. madrasensis* (6 per cent) and *S. cucullata* (4 per cent).

Medha Creek

Medha creek in Miani is about 100 m in length, upstream portion is delimited by an irrigation project shutter to prevent saline intrusion upstream. The saline condition persists throughout the year and abundant biomass of edible oysters and clams are observed. The average depth of the creek is about 2 m and good attachments of oysters (81 mm) were noticed. The fishermen go to the oyster beds by canoe during low tide and the oysters are detached from the rocks with a chisel.

Chemical Analysis

The edible oysters were represented by *C. gryphoides* (69 per cent), *C. rivularis* (11 per cent), *C. madrasensis* (8 per cent) and *S. cucullata* (12 per cent) in the order of abundance. There is no regular organized fishery as such for oysters but it is usually harvested in December – January for local consumption.

Scientific Name

Arca inaequivalvis, Bruguiere

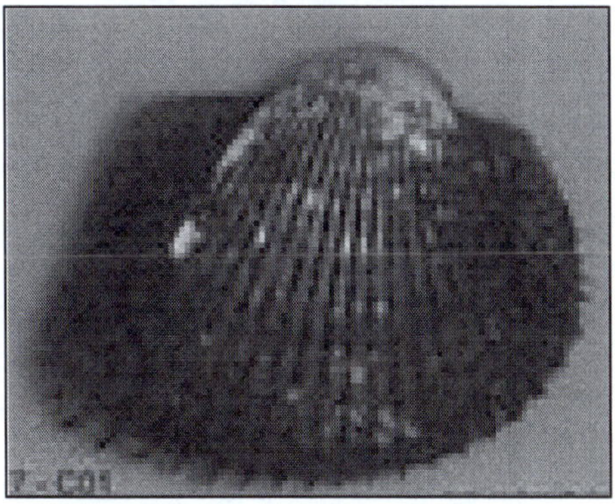

Classification

Phylum	:	Mollusca
Class	:	Bilvia
Sub Class	:	Fillibranchia
Order	:	Pteriomorpha
Super Family	:	Arcoidea
Family	:	Arcidae
Genus	:	Arca
Species	:	*inaquivalvis*

Reference

Arca inaquivalvis, Bruguiere, Encyclopedie, Methodique, Vers. I, 1792,

The shell is moderately large and only slightly longer than high, the height being definitely greater in proportion to the length than in most of the preceding species. The surface is strongly radiate ribbed, the ribs being broadly flattened, smooth, and numbering about thirty three. The shell is whitish, sometimes tinged with yellowish brown or orange, and often secondarily stained greyish black. The shell being rather thin, the external ribs often show through on the inner surface also.

Scientific Name

Acra granosa Linnaeus, 1758

English Name

Ark Shell

Synonyms

Anadara granosa kamakurensis Noda, 1966

Classification

Phylum	:	Mollusca
Class	:	Bilvia
Sub Class	:	Fillibranchia
Order	:	Pteriomorpha
Super Family	:	Arcoidea
Family	:	Arcidae
Genus	:	Arca
Species	:	*granosa*

Shell equivalve, thick and solid, ovate, strongly inflated, slightly longer than high and feebly inequilateral. It is about 18 radial ribs (15 to 20) with wide interstices at each valve. Ribs stout and distinctly rogues, bearing regular, often rectangular

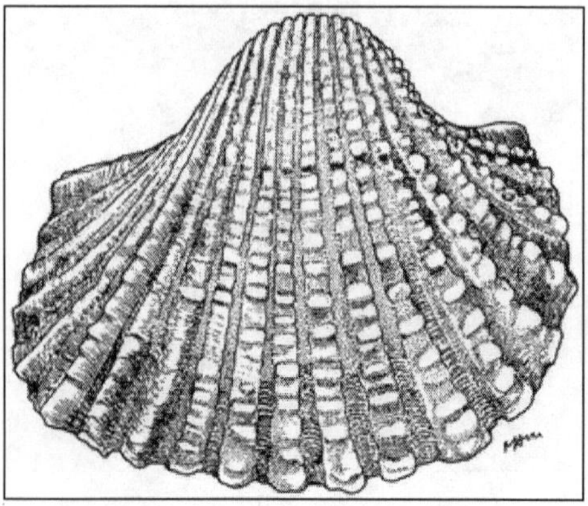

nodules. Periostracum is rather thin and smooth. Internal margins with strong crenulations corresponding with the external radial ribs. No byssal gape. Outside of shell white under the yellowish brown periostracum. Inner side white often tinged light yellow towards the umbonal cavity.

Scientific Name

Anadara granosa

Classification

Phylum	:	Mollusca
Class	:	Bilvia
Sub Class	:	Fillibranchia
Order	:	Pteriomorpha
Super Family	:	Arcoidea
Family	:	Arcidae
Genus	:	Anadara
Species	:	*granosa*

Anadara granosa is a species of ark clam known as the blood cockle due to the red haemoglobin liquid inside. It is found throughout the Indo-Pacific region from eastern Africa to Australia to Polynesia to Japan. It lives mainly in the intertidal zone at one to two metres water depth, burrowed down into sand or mud. Adult size is about 5 to 6 cm long and 4 to 5 cm wide. It has a high economic value as food, and it is kept in aquaculture. It is served steamed, boiled, roasted, or traditionally raw.

Scientific Name

Barbatia candida Helbling

Classification

Phylum	:	Mollusca
Class	:	Bilvia

Order : Arcoida

Family : Arcidae

Genus : Barbatia

Species : *candida*

The shell is 1 1/2 to 2 1/2 inches in length, fairly thin not heavily. Irregular in shape, byssal opening is at base of shell. Numerous weak, slightly beaded ribs, those on the posterior dorsal area being bery strongly beaded. Periestracum is brown, longest at posterior end. Exterior and interior of shell white. Ligament moderately developed.

Scientific Name

Barbatia domingensis Lamarck

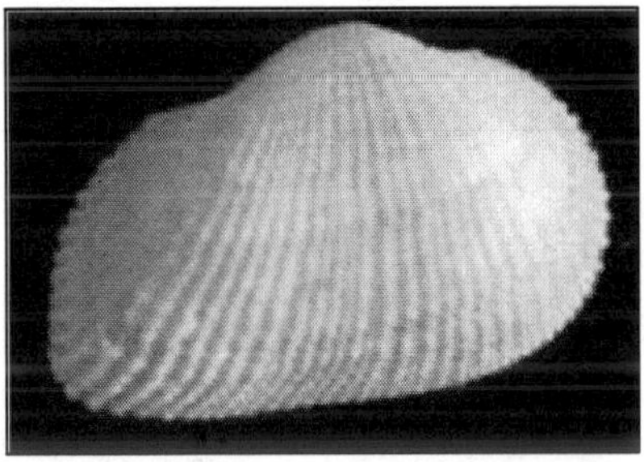

Classification

Phylum : Mollusca

Class : Bilvia

Sub Class : Pteriomorphia

Order : Arcoida

Family : Arcidae

Genus : Barbatia

Species : *dimingensis*

The shell is 1/2 to 3/4 inch in length, somewhat box shaped, and whitish in colour and with no appreciable perio stratum. It is similar in shape and scripture to *Acropsis adamsi*, but instead of having a small triangular. Ligament between the beaks, *domingensis* have a very narrow, long ligament posterior to the beaks. The posterior end is usually larger than the anterior end. Characteristically dip slightly downward. It is very common at low tide under rocks.

Scientific Name

Barbatia tensera

Classification

Phylum : Mollusca

Class : Bilvia

Sub Class : Pteriomorphia

Order : Arcoida

Family : Arcidae

Genus : Barbatia

Species : *tensera*

The shell is 1 to 1/2 inch in length, thin-shelled, rather far and evenly trapezoidal in shape and with numerous rather evenly and finely beaded, thread like ribs ligamental area fairly wide at the beak end, becoming narrow at the other. Small abyssal gape present on the ventral margin. They are common in Gulf area.

Scientific Name

Cardium setosum Red fern

Classification

Phylum : Mollusca

Class : Bilvia

Sub Class : Heterodanta

Order : Veneroida

Super Family : Cardiacea

Family : Cardiidae

Sub-Family : Cardiinae

Genus : Cardium

Species : *setosum*

Synonyms

Cardium latum, Chemnitz

Cardium setosu Widalgo

Reference

Cardium setosum Red fern, *Ann. Lyecum. Nat Hist.* New York, IV 1848, P 168

The shell is white, transversely oblong, with the longest dorso–ventral axis running obliquently from the umbo towards the hind part of the ventral margin. The ribs number about forty. They are markedly flattened, very closely set, and each of them bears a fins, median longitudinal groove from which arises arrow of spined which are found intact only in fresh specimens, they are generally reduced to small granules in partially worn shells, and are observed to persist longest towards the margins, but in very old and greatly worn out shells they completely disappear. The rins are separated by fins, thread like grooves, which are well developed, the anterior lateral tooth being specially large and prominent. The posterior lateral tooth is removed a considerable distance away from the umbo. The interior of the shell is pure white, smooth and highly polished, but the margin of the inner surface is conspicuously cut up into a series of squarish, truncated teeth corresponding to the terminations of the external radial grooves. Externally the shell is white, sometimes tinged with pale yellow.

Scientific Name

Caridta bicolour, Hornel

Classification

Phylum	:	Mollusca
Class	:	Bivalvia
Sub Class	:	Heterodanta
Order	:	Veneroida
Super Family	:	Carditacea
Family	:	Cardiidae
Sub-Family	:	Cardiinae
Genus	:	Cardita
Species	:	*bicolor*

Synonyms

Acetnobolus antiquates, Admas

Venericardia antiauata, Smith.

Reference

Cardita bicolour, Hornell, *Common molluscs of South India, Mad. Fish. Bull., XIV*, 1921, P. 177

The shell is thick, solid and bears a superficial resemblance to a cardium. The valves are fairly deep and sore what obliques, *i.e.* the largest dorso-ventral axis of the shell runs obliquely from the umbo to the ventral margin. The umbo is rather conspicuously curved anteriorly. There are two elongated posterior teeth extending from the umbo and separated from each other by a deep groove. The anterior tooth is minute and some time almost absent. The shell is sculptured with flattened surfaces. The shell is white, marked with brownish or bright orange spots which are mainly confined to the surfaces of the ribs and tend to be larger interiorly. The inner surface is white and its margin appears strongly toothed owing to the presence of the external radial ribs.

Scientific Name

Chione fluctifraga Sowerby

Classification

Phylum	:	Mollusca
Class	:	Bivalvia
Sub Class	:	Heterodanta
Order	:	Veneroida
Family	:	Veneridae
Genus	:	Chione
Species	:	*fluctifraga*

Synonyms

Chionista fluctifraga

The shell is 1½ inch in length, slightly longer, moderately compressed, sub triangle, radial grooves or ribs strong at the posterior third and at the anterior quarter of the shell, central area with stronger, low rather wide, concentric ribs which may have courses, half moon shaped beads. Lingual not well defined, smutch creamy white, semi glossy, rarely stained with blue grey. Interior is with purple splotches near the muscle scars or on the teeth.

Scientific Name

Chlamys sentis Reeve

Classification

Phylum	:	Mollusca
Class	:	Bivalvia
Sub Class	:	Metabranchia
Order	:	Pterionorpha
Super Family	:	Pectinoidea
Family	:	Pectinidae
Sub-Family	:	Chlamydiae
Genus	:	Chlamys
Species	:	*Sentis*

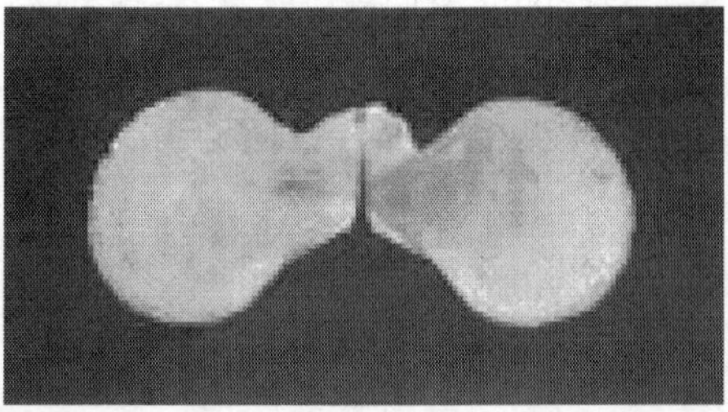

The shell is 1 to ½ inches in length, but not so wide (like a fan opened only 8-degrees) valves rather flat one hinge ear small, the other twice as large with about 50 ribs of varying sizes, each with tiny, closely set scales. There are 2 to 4 smaller ribs between the slightly larger ones. Colour commonly brilliant purple red, vermilion, orange-red, brownish, white or mottled (especially near the beaks). It is common under rocks below low tide mark.

Scientific Name

Crassostrea gryphoides Schlothem

Classification

Phylum	:	Mollusca
Class	:	Bivalvia
Sub Class	:	Palaeocheterodonta
Super Family	:	Crassistreacea
Family	:	Crassostridae
Genus	:	Crassostrea
Species	:	*gryphoides*

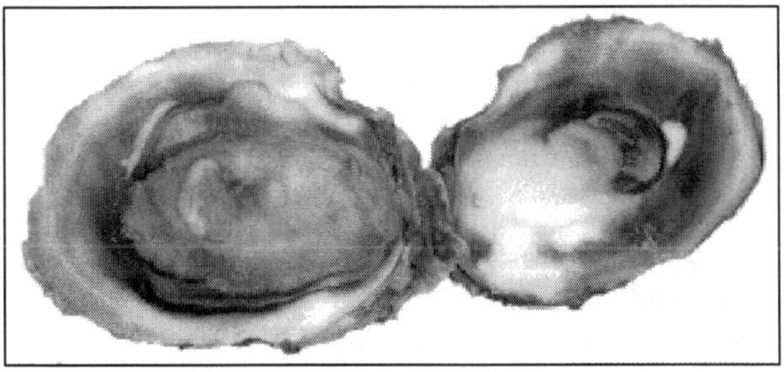

The soft body of the animal is encased by two valve – lower cupped valve (left valve) and upper flat right valve. While the lower valve is commented to substratum of the upper flat valve acts as lid to open and close by hinge mechanism connecting both the valves. The shell consists of three layers. The upper most layers are known as peristracum. The inner layer is the nacreous layer which is thin, hard and usually shiny. The middle layer is thick and chalky. The shell is primarily composed of calcium carbonate. The peristacum is almost all protein. The texture size, shape, marking and colour of the shell vary between species and from place to place. The same species grown in different environment exhibit different in shell characters. Animal mantle is pigment black. Anal portion of the rectum is 1 to 2 mm in length and markedly directed outwards. The anal tip is simple, at times slightly funnel shaped. Anal opening situated at the corner of the posterior and ventral margins of the adductor muscle. Adductor muscle is roundest, bean shaped rarely oblong and elliptical (Dholakia, 2004).

Scientific Name

Crassostrea cucullata, Rao, 1969

Classification

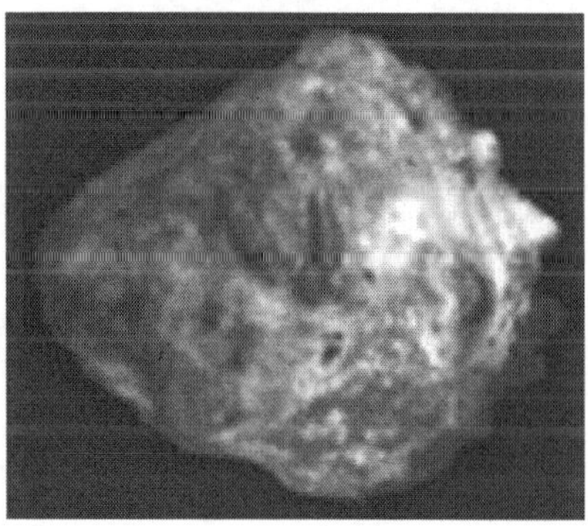

Phylum	:	Mollusca
Class	:	Bivalvia
Sub Class	:	Palaeocheterodonta
Super Family	:	Crassistreacea
Family	:	Crassostridae
Genus	:	Crassostrea (saccostrea)
Species	:	*cucullata*

Synonyms

Ostrea forskali, Chemnitz 1785.

Ostrea cucullata Born, 1780

Saccostrea cucullata Born

Reference

Crassostrea cucullata, Rao, 1969, *Indian Farming* 29 (9): 41

The shell valves of this species are hard and stony and the shape is trigonal or pear shaped. The left valve is deep or moderately so. The right valve is flat or slightly convex and covers the left one like algid. The hinge is straight, of moderate size and devoid of teeth and the umbonal cavity well developed. The margin of both valves has well developed angular folds sculptured with laminate. Small turborcles present along the inner margin of the right valve and there are corresponding pits in left valve. Adductor scar is kidney shaped, straitened and whiter or grayish in colour. Colour of the outer surface of valve is variable being pale white, grey light or dark brown, green or purplish In India it occurs at various places along both the east and west coasts and around Andaman and Lakshadweep Island, In Gujarat Armda, Dwarka, Adatra and Hanuman Dandi (Dholakia, 2004).

Scientific Name

Crassostrea madrasensis, Rao

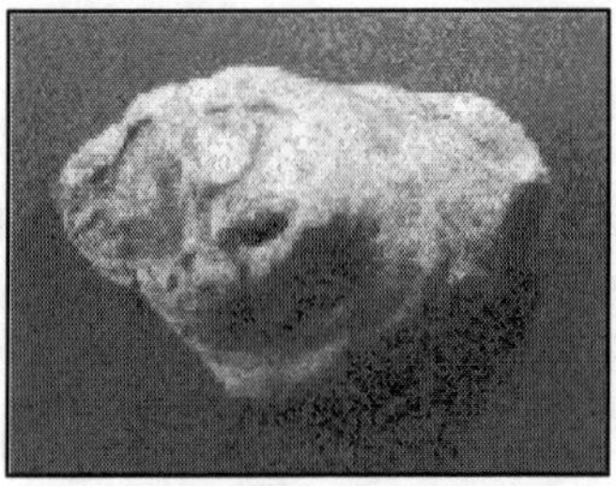

Classification

Phylum	:	Mollusca
Class	:	Bivalvia
Sub Class	:	Paleonoterodonta
Super Family	:	Crassostreacea
Family	:	Crassostreidae
Genus	:	Crassostrea
Species	:	*madrasensis*

Reference

Crassostrea madrasensis, Rao, 1974, Chapt. II, In the Commercial Molluscus of India, *Bull. Cent. Mar, Fish.*

The shell valves are very irregular in shaped. They are usually elongated. When spat set on flat surfaces and there is no elongate. When spat flat surface and there is no crowding a flat shape of the niche where they are present and over crowding leads to oysters with very much twisted shells. This species grows to a size of 212 mm,. the larger ones occurring in estuarine systems. It is a typical euryhaline species and flourishes especially well in turbid brackish water like estuaries, creaks, bays and backwaters growing to a large size with heavy meat. It is also found in sheltered areas like ports and harbor where it occurs in large numbers attached to pillars, walls of wharfs and buoys and along open coasts. It is found from mid littoral zone to a depth of 15 – 16 meters.

Scientific Name

Crassostrea cristagalli Rao, 1974

Classification

Phylum	:	Mollusca
Class	:	Bivalvia
Sub Class	:	Paleonoterodonta
Super Family	:	Crassostreacea
Family	:	Crassostreidae
Genus	:	Crassostrea
Species	:	*cristagalli*

Synonyms

Ostrea cristagalli Linnasus

Reference

Crassostrea cristagalli Rao, Ch. II. In the Commercial Molluscs of India, *Bull. Cent. Mar. Fish. Res. Inst.*, 25: 36 - 37.

The shell is broadly rounded or sub-quadrate in shape with the margin of valves thrown into deep, sharp, angular folds and interlocking with each other. The outer surface of the folded margins has diverging, closely set, granulated striate. The external shell colour varies from brownish to violet and the internal surface of the valve is grayish white. It occurs attached to stones or dead corals and grows to a size of 60 mm. At Poshitra point, Harshad areas, Gujarat bed of the species are found and the clusters are collected and consumed by fisher folk locally. Stray individual of the species occur along the Thanjavur coast, Palk Bay and Gulf of Mannar.

Scientific Name

Crassostrea folium Rao. 1974

Classification

Phylum	:	Mollusca
Class	:	Bivalvia
Sub Class	:	Paleonoterodonta
Super Family	:	Crassostreacea
Family	:	Crassostreidae
Genus	:	Crassostrea
Species	:	*folium*

Synonyms

Ostrea folium Chemnitz 1784.

Reference

Crassostrea folium, Rao,1974, Ch-II. In the Commercial Molluscs of India, *Bull. Cent. Mar. Fish. Res. India* 25: 37

Shell broadly crate in shape. The left valve is deeply concavely excavated along the middle to fit surface of attachment. Right valve is raised into a broad tube-like longitudinal rib in the middle corresponding to the excavation of left valve. Shell has a number of rounded folds diverging away from the rib more or less smooth except for a few thin overlapping laminate towards the margin. Colour of valves pale grey or brownish purple externally and grayish internally. It is found in Positra area in Gulf of Kutch. It is not of economic value. It grows to a size of about 50 mm.

Scientific Name

Dinocardium robustum Solander

Classification

Phylum : Mollusca

Class : Bivalvia

Sub Class : Heterodonta

Order : Venerodia

Super Family : Cardioidea

Family : Cardiidae

Genus : Dinocardium

Species : *robustum*

The shell is 3 to 4 inches in size, ovate, inflated, with 32 to 36 rounded, radial smoothies ribs. Externally saw yellow with its posterior slope mahogany red shading toward purple near the edge. Interior rose, with brownish posteriorly and with a white anterior margin. This is the large, common in Gulf of Kutch near Narara reef., Poshitra area. It is also big size seen in the Gulf of Kutch area.

Scientific Name

Donex cuneatue Linne

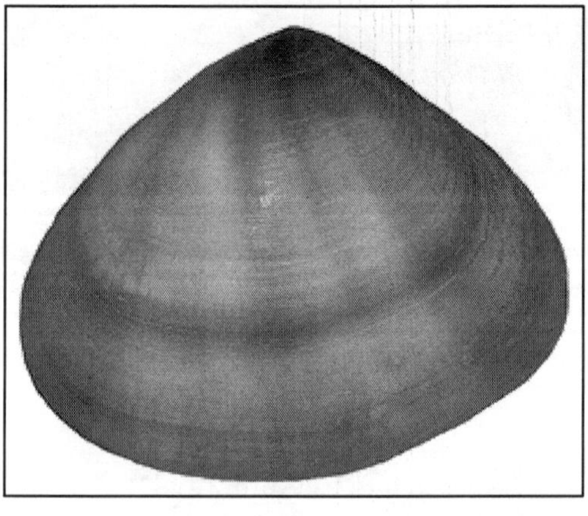

Classification

Phylum : Mollusca

Class : Bivalvia

Sub Class : Heterodonta

Order : Venerodia

Super Family : Teltinidae

Family : Dinacidae

Genus : Donex

Species : *cuneatue*

Synonyms

Donax cuneata, Linne

Reference

Donex cuneatus, Gravely, *Bull. Mad. Govt. Mas. (Nst.Hist).* V, No 1, 1941

The shell is moderately large, thick, rather compressed, inequilaters and elongately triangular, the posterior side being much shorter and inclining more steeply down from unbolt than the anterior side. This area appears rough and coarsely granulated on casual observation. The rest of the surface is finely radiantly striated. The inner surface is smooth and its margin untoothed. The colour of the shell is subject to considerable variation. It is generally pale purplish or bluish grey. Marked with fine, irregular, reddish brown radial lines, somewhat interrupted and often branching or coalescing.

Scientific Name

Donax variabilies

Classification

Phylum	:	Mollusca
Class	:	Bivalvia
Sub Class	:	Heterodonta
Order	:	Venerodia
Super Family	:	Teltinidae
Family	:	Dinacidae
Genus	:	Donex
Species	:	*variabilies*

Shell is medium to small size (1/2 to 3/4 inches). The thin, transparent perio – stratum gives a gloss to the valve. The colouration is externally variable, white, yellow, pink, reddish, and violet usually with ray of deeper colour. The inner surface shows the same range of outer surface. A well developed mobile foot, a good digger is found on body. The two lobes of the mantle are fused, learing an aperture at the anterior for the foot, and at the posterior they are extended. Each gill consists of two parries of lamellas extensively fused by inter lamellar junctions. Their habitat is soft bottom, usually sandy in the intra-littoral zone.

Scientific Name

Gafrarium dispar Chemnitz

Classification

Phylum	:	Mollusca
Class	:	Bivalvia
Sub Class	:	Heterodonta
Order	:	Venerodia
Family	:	Veneridae
Genus	:	Gafrarium
Species	:	*dispar*

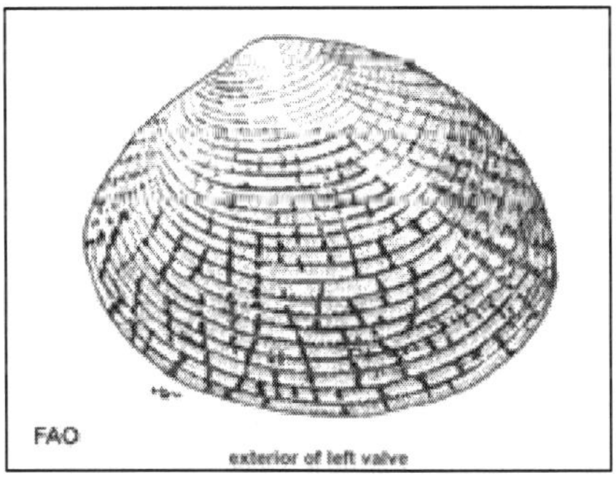

Synonyms

Venus dispar, Chemnitz

Circe disper, Fisher

Cytherea dispar, Catlow and Reeve.

Reference

Gafrarium dispar, Chemnitz, *Conch, Cab.*, XI, 1795, P. 230

The shell of this species is easily distinguished, is much weaker, consisting of moderately close set, concentric ridges and oblique, divaricating radial ridges, the latter being conspicuous only in the posterior part of the shell. The anterior and posterior margin of the shells is more or less similarly rounded and the umbo is more centrally placed than in the preceding species. The hinges teeth are thinner and slightly more narrowly diverging. The lunule is well defined, elliptical and usually of a uniform dark brown colour. The area behind the umbo is large, steeply sloping inwards and marked with widely spaced, purplish, transverse lines. The outer surface is of a pale yellowish white colour, dotted all over with orange red or brownish spots, some of which sometimes tend to coalesce forming zig-zag marking.

Scientific Name

Grafrarium divaricata Chemnitz 1782

Classification

Phylum	:	Mollusca
Class	:	Bivalvia
Sub Class	:	Heterodonta
Order	:	Venerodia
Family	:	Veneridae
Genus	:	Gafrarium
Species	:	*divaricata*

FAO

Synonyms

Cytherea divaricata, Lamarck

Crista divaricata, Dautsenberg and Fishcher

Reference

Gafrarium divaricata Chemnitz, *Conch, cab.* VI, 1782, P. 317 Pl XXX.

The shell is larger and thicker than in any of the preceding species of *Grararium*, but with a relatively weak sculpture, consisting of concentric and divaricating radial ridges, burin this species the later are developed uniformly both anterior and posterior to the line of divaricating. The concentric ridges however, tend to lose their distinctness posterior. There are also a few specially well marked concentric growth lines at somewhat wide intervals. The outer surface is pale yellowish or whitish, profusely ornamented with reddish brown angular marking with their paces turned towards the umbo. This pattern is often supported by brownish radial lines extending inwards to a little distance from the margins. The inner surface is smooth and white. The margin of the shell is thick, and in the adult, often consists of several supreme posed growth layers.

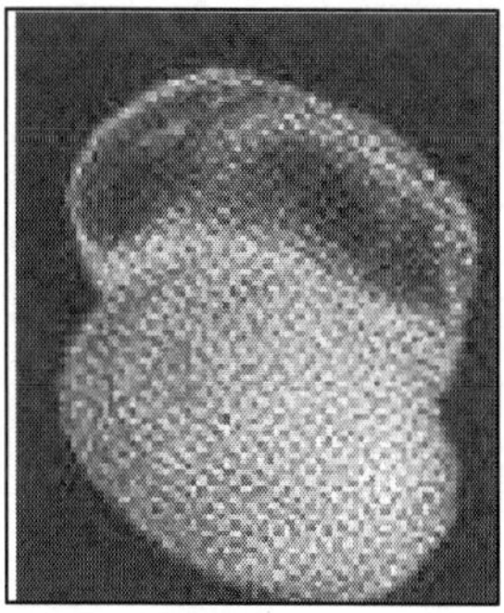

Scientific Name

Irus lamellifer Conred

Classification

Phylum	:	Mollusca
Class	:	Bivalvia
Sub Class	:	Heterodonta
Order	:	Venerodia
Super Family	:	Veneroidea
Family	:	Veneridae
Genus	:	Irus
Species	:	*lamellifer*

The shell is 1 to 1½ inches in length, usually oblong, although some specimens may be almost round. Characterized by about a dozen, is strongly raised, concentric lamellae or thin ridges. Shell whitish and with a chalky texture. Moderately common and found burrowing in Gray shale from low water and several fathoms.

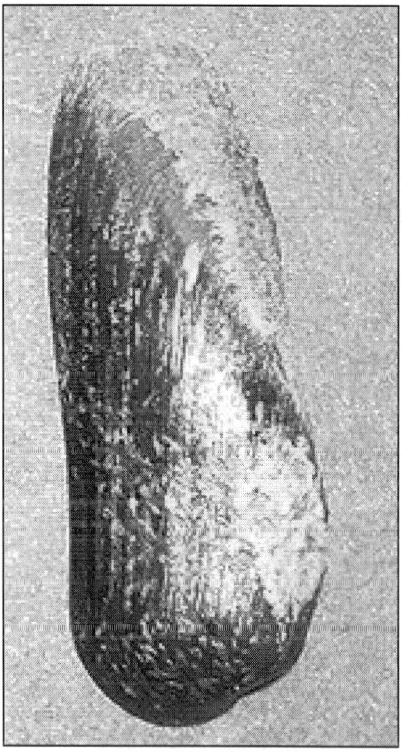

Scientific Name

Lithophaga antillarum Orbigny

Classification

Phylum	:	Mollusca
Class	:	Bivalvia
Order	:	Mytiloida
Family	:	Mytilidae
Genus	:	Lithophaga
Species	:	*antillarum*

The shell is 2 to 4 inch in length, elongate, cylindrical and colour red light yellowish brown on the outside and iridescent cream inside. Sides of valve marked with numerous, irregular, vertical rib lets. They are fairly common in soft rocks in moderately deep water.

Scientific Name

Mactra turgida Gray

Classification

Phylum : Mollusca

Class : Bivalvia

Sub Class : Heterodonta

Order : Veneroida

Super Family : Mactracea

Family : Mactridae

Sub Family : Mactrinae

Genus : Mactra

Species : *turgida*

Reference

Mactra turgida, Gray, *Mag. Nat Hist. Ms., I* 1837, P. 371

Synonyms

Mactra turgida Gamlin, Syst. Nat. Ed. XIII, 1790, P. 326

The shell is large, strongly inflated rather thick or its size, whitish and is easily recognized by its smooth, glossy surface and prominent, elevated umbo which is often tinted pinkish violet. The shell is triangularly oval, almost equilaterally, with the front and hind margin about equally broad and rounded. The surface of the shell is nearly entirely smooth and polished, but fire irregularly spaced on the lunule and the area behind the umbo. The umbo is conspicuously elevated, and the umbonal area and the part of the surface immediately below it are smooth and highly polished. The pit for the ligamentary nodule is relatively larger in this species and the lateral teeth are longer, the anterior being more so. The pallial sinus is short,broad, obliquely inclined upwards and some what sqarish at its apex. The anterior surface is smooth and white, and the outer surface is generally white over the great part, but occasionally yellowish white or pale grayish, and almost always coloured deep pinkish violet towards the umbo.

Scientific Name

Mercenaria campechiensis Gmelin 1791

Classification

Phylum : Mollusca

Class : Bivalvia

Sub Class : Paleotaxodonta

Order : Veneroidae

Super Family : Veneracea

Family : Veneridae

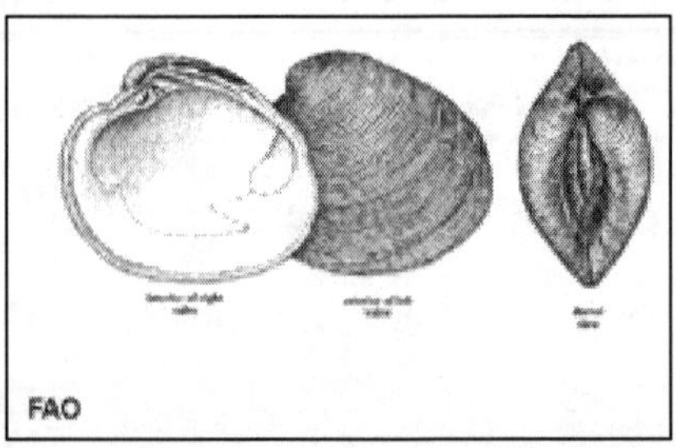

Sub Family : Venerinae

Genus : Mercenaria

Species : *campechiensis*

The shell is 3 to 6 inches in length, very similar to *mercenaria*, but much more obese, heavier, shell, lacks the smooth central area on the outside of the valves and the entire lunule is usually as wide as long. Always white infernally rarely it has a purplish stain on the escutcheon and brown mottling on the side. There have been a number of forms described. In the vicinity of Florida, there is malformed race in which there is a sharp elevated ridges passing from the umbo obliquely backward towards the pallial sinus on the inside of each valve.

Scientific Name

Mercenaria mercenaria Linne

Classification

Phylum : Mollusca

Class : Bivalvia

Sub Class : Paleotaxodonta

Order : Veneroidae

Super Family : Veneracea

Family : Veneridae

Sub Family : Venerinae

Genus : Venus

Species : *mercenaria*

General size of this mollusc is about 3 to 5 inches in length, ovate-tri-gonal, about 5/6 as high, heavy and quite thick, moderately inflated. Sculpture of shell is numerous, concentric lines of growth or small rebels. Near the beaks these lines are prominent and distantly spaced. The exterior center of the valves has a characteristic smoothics or glossy area. Exterior part is dirty gray to whitish, interior white, commonly with purple staining. The entire lunule is 3/4 as wide as long. The form notate say from the same region is externally marked with brown, zigzag chowders clams on the half shell or "Cherrystones". It is also known as the Hard-shelled clam. It is characterized by a glossy central area on the outer side of the shell, but has larger irregular, coalescing, flat-topped, concentric ribs. Formally placed in the genus Venus many years ago, but almost universally placed in a genus by itself by modern workers.

The hard clam, burrows shallowly in sediments of either mud or sand. It is among the most commercially important species of invertebrate. Like other clams, it is filter feeder.

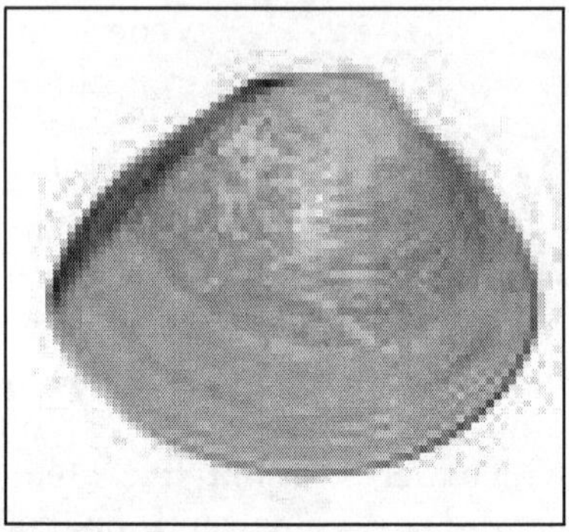

Scientific Name

Meretrix meretrix Gemelin 1791

Classification

Phylum	:	Mollusca
Class	:	Bivalvia
Sub Class	:	Paleotaxodonta
Order	:	Veneroidae
Super Family	:	Veneracea
Family	:	Veneridae
Sub Family	:	Venerinae
Genus	:	Meretrix
Species	:	*Meretrix*

The Asiatic hard clam Meretrix meretrix (Bivalvia: Veneridae) is a commercially important species inhabiting sand flats of Hong Kong and Southeast Asia. Its size is up to 90 mm in shell length and it is easily distinguished from other species.

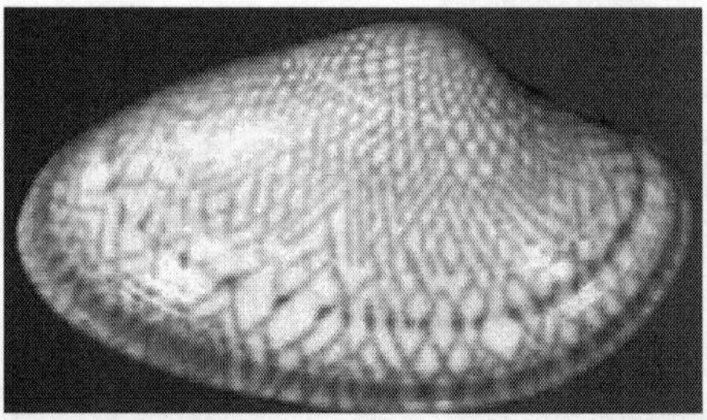

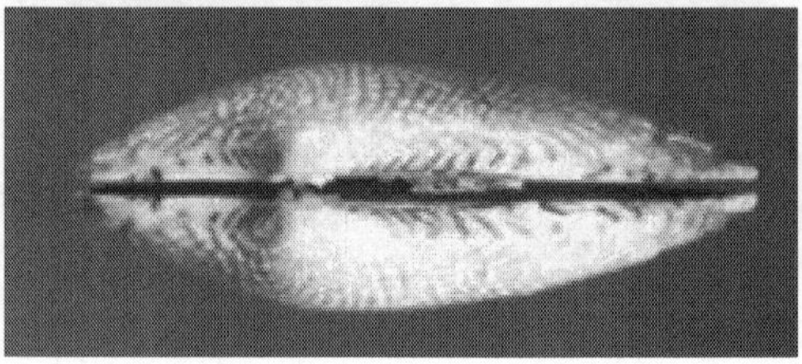

Scientific Name

Paphia textile Gmelin

Local/English Name

Clam

Synonyms

Venus textile Gmelin, 1791

Classification

Phylum	:	Mollusca
Class	:	Bivalvia
Sub Class	:	Heterodonta
Order	:	Veneroida
Family	:	Veneridae
Genus	:	Paphia
Species	:	*Textile*

Shell is medium sized, solid, equivalve, in equilateral; ambones well in the anterior part of shell. Outline elongates, almost twice longer than high; dorsal margin sloping to a narrowed posterior margin; anterior rounded slightly acute. Hinge without lateral teeth. Well defined lunule slightly depressed. Surface is shining. Sculpture smooth with growth lines only. Pallial sinus is round pointing to the beaks reaching half the shell's height. Internal margin is smooth. They are filter feeder. Burrows in deep continental shelf waters bottom where it can dominate the molluscan fauna. Color is externally beige, pale yellow with a characteristic zig-zag pattern in brown. Internally white common size: shell to 50 mm.

Scientific Name

Paphia malabarica Chemnitz

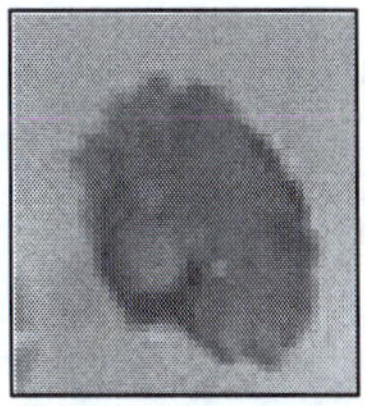

English Name

Yellow Clam

Classification

Phylum	:	Mollusca
Class	:	Bivalvia
Sub Class	:	Heterodonta
Order	:	Veneroida
Family	:	Veneridae
Genus	:	Paphia
Species	:	*malabarica*

Shell is subtrigonal, thick, solid, compressed, porcellaneous, smooth. Hinge strong, resilium narrow, pallial sinus short, white in colour. Ventral margin is light brown to chocolate. They are filter feeder. Burrows in deep continental shelf waters bottom where it can dominate the molluscan fauna.

Scientific Name

Pecten crasicostatus Sowerby, 1842

Classification

Phylum	:	Mollusca
Class	:	Bivalvia
Sub Class	:	Metabranchia
Order	:	Pteriomorpha
Super Family	:	Pectinoidea
Family	:	Pectinidae
Sub Family	:	Pectininae
Genus	:	Pecten
Species	:	*crassicostatus*

Reference

Pecten crassicostatus, Sowerby, Thes. Conchyl,. I 1842, P. 75

The valves are equilateral, rounded and about as high as long. The right valve considerably more flattened than the left. The surface of the ribs bears faint transverse curved striation throughout their length. The outer surface is bright reddish brow, marked all over with irregular white spots and blotches. The interior is glossy, whitish and smooth for the soft part, but towards the periphery there is a broad marginal fleshy pink band, this part being conspicuously cut up into broad, flattened, tooth like ridge by deep grooves which correspond in position with the external ribs.

Scientific Name

Pecten diegensis Dall 1898

Classification

Phylum	:	Mollusca
Class	:	Bivalvia
Sub Class	:	Metabranchia
Order	:	Pteriomorpha
Super Family	:	Pectiniodae
Family	:	Pectinidae
Genus	:	Pecten
Species	:	*Diegensis*

The shell is 2 to 3 inches in size. Right valve is convex with 22 or 23 flat topped ribs, which are generally longitudinally ridged on top. Left valve much flatter with 21 to 22 narrow, rounded ribs. Dredged from 10 to 75 fathoms.

Scientific Name

Pinctada fucata Gould

Classification

Phylum	:	Mollusca
Class	:	Bivalvia
Order	:	Pterioida
Family	:	Petriidae
Genus	:	Pictada
Species	:	*fucata*

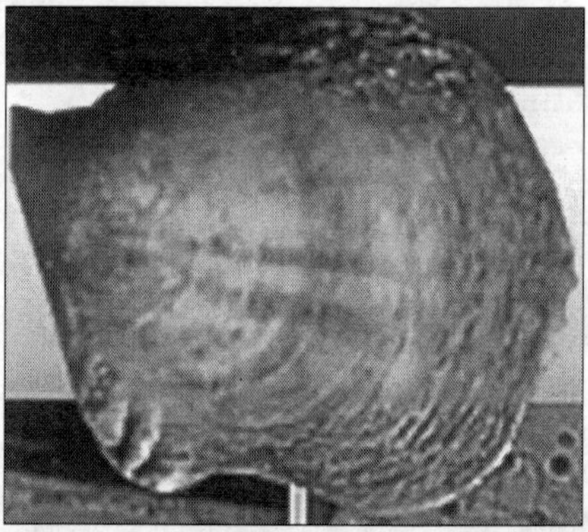

Pictada fucata

Synonyms

Pteria vulgaris Jameson 1901

Margaritifera vulgaris hornel, 1922

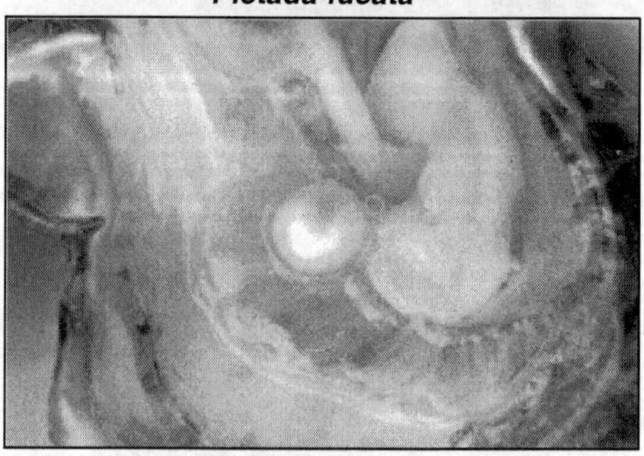

Oyster with pearl and mental

The hinge is fairly long, its ratio to the broadest region of the body of the shell is about 0.85 and its ratio to the longest dorso-ventral measurement is about 0.76. In both the valves there are hinge teeth, one at the anterior and posterior ends of the ligament. The anterior ear is larger than in other species and the byssal notch at the junction of the body of the shell and the ear is slit like; the posterior ear is fairly well developed. The posterior border of the shell shows a small or moderately large sinus. The adductor impression is large and sub central. There are 12 to 15 small scars caused by the insertion of pallial muscle between the umbo and the antero-ventral border.

The soft body of the pearl oyster consists of a viscera-pedal mass covered by the right and left mantle lobes which are free interiorly, ventrally and posteriorly but fused dorsally. The mantle edge has two thin, folds with pigmented, papillate edges. The outer fold is parallel to the inner surface of the shell and the inner fold called the pallial veil or velum projects at tight angle from the mantle edge. In life the pallial is of the opposite sides are in contact with each other except in the regions of the inhalant aperture about min-ventrally and the exhalent aperture at the posterior end. The foot is elongated and muscular arising about mid-way between the mouth and the intestinal lone and has a grove ventrally. The byssus is at the proximal end of the foot and bears byssus fibers with which the pearl oyster attaches itself to the substratum.

Anatomy of Pearl Oyster

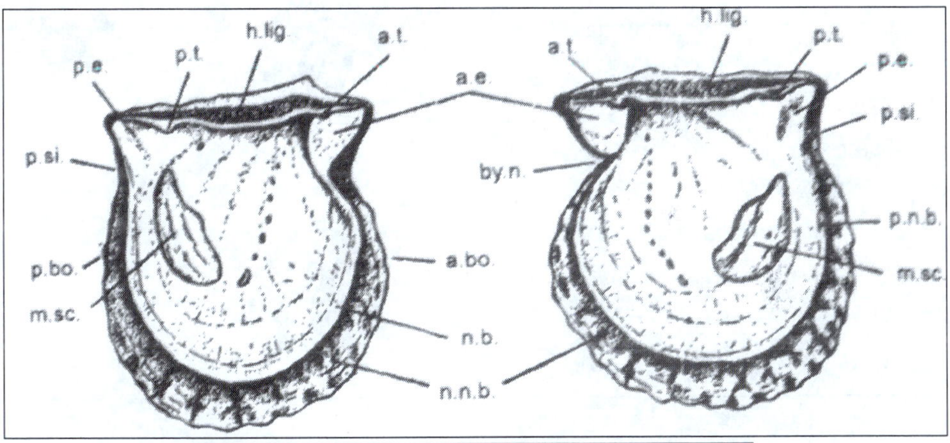

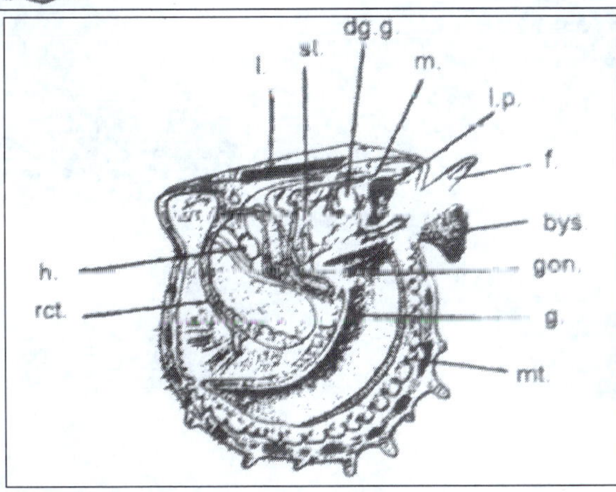

f: Foot; mt: Mantle; bys: Byssus thread; h: Hinge; st: Stomach; l.p. Libial palps; dg: Digestive gland; gon: Gonad; p. bo: Pallial border; m.sc: Muscle scar

Importance of Pearl Oyster in India

In India Pearl Oysters are available only at two places, one in Gulf of Kutch (Gujarat), and second in Gulf of Mannar, (Near Andaman Nicobar and Tuticorin). Govt. of India has declared this as endangered species and has to be protected. The author has produced 35 lakhs of juveniles after breeding them in the laboratory at Sikka and released them in Gulf of Kutch to increase population and natural stock (Dholakia, 2004).

Scientific Name

Perna viridis (Linnaeus, 1758)

Common Name

Asian Green Mussel

Classification

Phylum	:	Mollusca
Class	:	Bovalvia
Sub Class	:	Pteriomorpha
Order	:	Pectiniodae
Family	:	Mytilidae
Genus	:	Perna
Species	:	*Viridis*

Synonyms

Chloromya viridis Dodge, 1952

Mytilus (Chloromya) smaragdinus Jukes-Browne, 1905

Mytilus (Chloromya) viridis Lamy, 1936

Mytilus opalus Lamarck, 1819

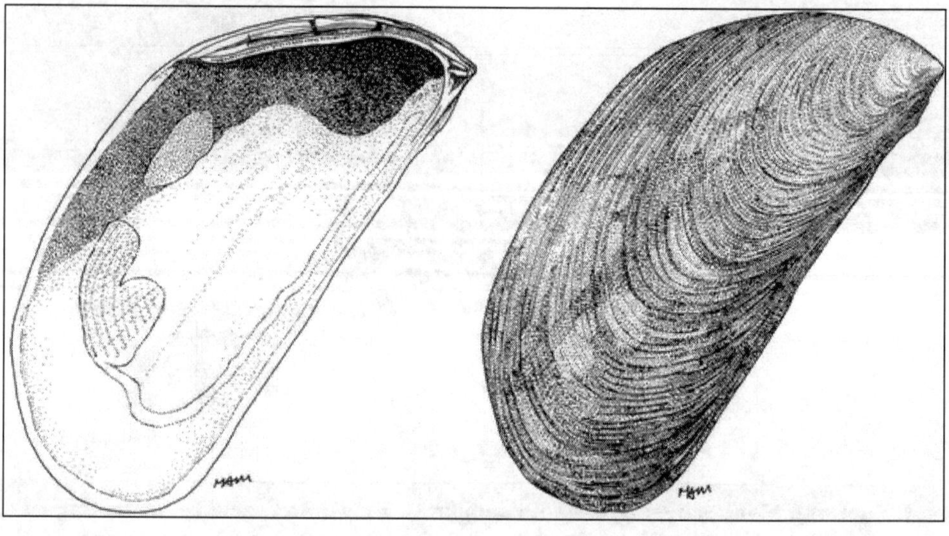

Mytilus smaragdinus Chemnitz, 1785

Mytilus viridis Linnaeus, 1758

Reference

Bivalvia By J. M. Poutiers

Perna viridis can reach 165 mm in length, but is usually 80-100 mm. It is a typical mussel shape, roughly trigonal-ovate in outline, swollen when viewed from above, with a sharp beak at the anterior end, and rounded and compressed posteriorly. The beaks are incurved (downturned) and the ventral margin is straight or weakly convex. The outer surface of the shell valves is nearly smooth and covered with periostracum (epidermis) that can be vivid green or dark brownish-green anteriorly, and olive-green posteriorly. However, the periostracum peels off when the animal dies and the shell dries out. The interior of the shell valves is shiny and iridescent (pearly) pale bluish green. The two shell valves are connected by a horny brown ligament and the ridge that supports this ligament is finely pitted. The beak has microscopic hinge teeth, one in the right valve and two in the left valve, that interlock when closed. Where the mantle (the part of the animal that secretes the shell) is attached internally is evident by a circular line (pallial line). The posterior end of this line is sinuous (wavy) a key feature of this species. The shell valves are closed shut by a large, kidney-shaped adductor muscle near the posterior end. Minimum Length: 15 mm, Maximum Length: 165 mm Sexually mature at 15-30mm shell length (2-3 months age).

Distribution

East and West coast of India and off Port Blair.

Areas of Abundance

Coasts of Kerala, Karnataka, Ratnagiri, Gulf of Kutch and Kakinada Bay.

Scientific Name

Pinna vexillum Bora

Classification

Phylum	:	Mollusca
Class	:	Bovalvia
Sub Class	:	Pteriomorpha
Order	:	Pectiniodae
Family	:	Pinnidae
Genus	:	Pinna
Species	:	*Vexillum*

Reference

Chimnitz viii p. 238 +91 F 783

Shell about five inch long and 3 broad of brownish horn colour with dusky rays; it has several longitudinal ribs, of which a few towards the middle of the shell are

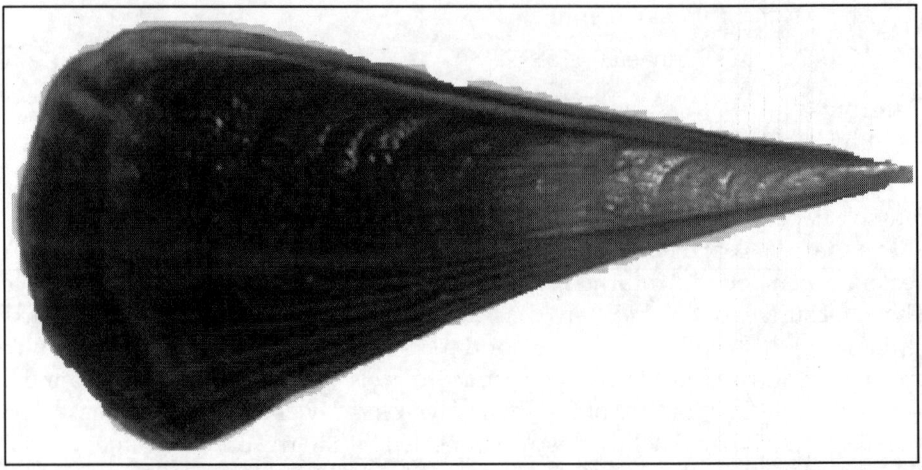

armed with vaulted scales, and it transversely wrinkled especially towards the margin. The shell is sub-triangular ovate with the posterior margin ditted; anterior side longitudinally ribbed and the other transversely wrinkled.

Scientific Name

Pinna bicolor Gmelin

Classification

Phylum	:	Mollusca
Class	:	Bovalvia
Sub Class	:	Pteriomorpha
Order	:	Pectiniodae
Family	:	Pinnidae
Genus	:	Pinna
Species	:	*bicolor*

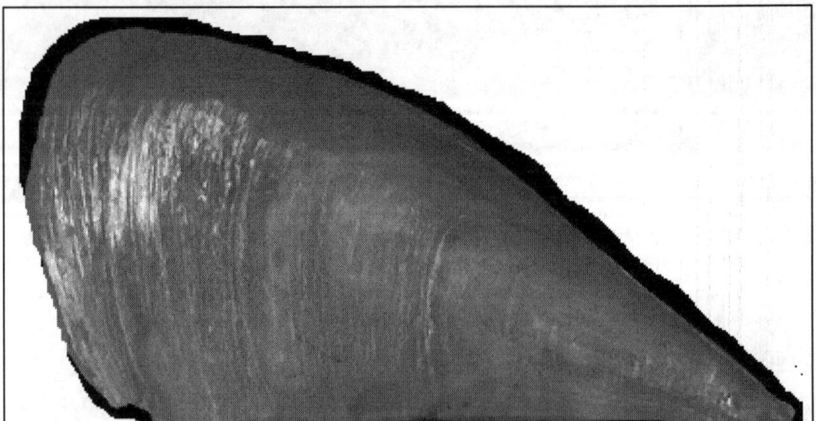

Reference

Chemnitz viii p. 234 + 90 f. 780

Shell about five inch long and 3 broad of brownish horn colour with dusky rays; it has several longitudinal ribs, of which a few towards the middle of the shell are armed with vaulted scales, and it transversely wrinkled especially towards the margin. Shell ovate oblong, incurved towards the summit and rather glabrous with only a few longitudinal striate.

Scientific Name

Pinna Nigra Dil

Classification

Phylum	:	Mollusca
Class	:	Bovalvia
Sub Class	:	Pteriomorpha
Order	:	Pectiniodae
Family	:	Pinnidae
Genus	:	Pinna
Species	:	*Nigra*

Reference

Chemnitz viii P. 221 + 88 f. 774

Shell is ovate, ventricose, with the margin rounded on both side and striated longitudinally. it has several longitudinal ribs, of which a few towards the middle of the shell are armed with vaulted scales, and it transversely wrinkled especially towards the margin.

Scientific Name

Placenta placenta Linne.

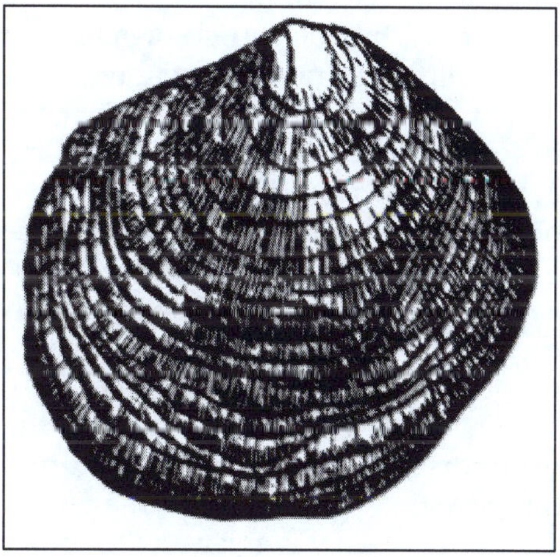

Local English Name

Window-pane oyster

Classification

Phylum	:	Mollusca
Class	:	Pelecypoda
Order	:	Eulamellibranchina
Sub Order	:	Ostracea
Family	:	Ostreidae
Genus	:	Placenta
Species	:	*Placenta*

Synonyms

Anomia placenta Linnaeus 1758.

Reference

Placenta placenta, Mossess, S.T. *J. of Natural Hist. Soc.* Vol XII.

The local name of this species is Window pane oyster. It is well known for obtaining mother of pearls and pearl dust valued at several lakhs of rupees per season in gulf of Kutch. No where else along the Indian coast, the window pane oyster fishery is of comparable importance, except at Kakinada Bay. This is one of the commercially important molluscan fisheries of Saurashtra Gujarat. It can be considered to be next only to the window pane oyster fishery of Kakinada Bay in India,

The shell of adult oyster is free, very much compressed, slightly unequal, sub-orbicular in shape, the height and length approximately equal. The shell valves are slightly unequal very flat, rounded and translucent. The inner surface of the valves is pearly. The adductor impression is at about center. The umbo is small. Two thin ridge like teeth diverge from the umbo making a characteristic inverted V shape angle. The shell is fairly large with a diameter of 14 cm or even more. It bears numerous concentric line of growth on the exterior consisting of slightly projecting lamellae the margins of which are minutely uneven with finger like or spatulate processes. The shell of the adult window pane oyster is white in colour.

The body is very much latterly compressed and is covered by two folds of mantle. Interiorly on either side of the visceral mass is a pair of elongated, narrow labial pulps between which is the mouth. There is a pair of gills on either side of the visceral mass. The foot is cylindrical, flattened laterally and attached to the anterior surface of the visceral mass and its tip has a deep cuplike sucker, which facilitates dispersal of foreign matter settling on the body. The alimentary canal consists of a slit like mouth, esophagus, a large stomach surrounded by the digestive gland, well developed pyloric caecum with crystalline style, a short intestine a rectum which ends by anus in the center of the broad ventricle membranous collar. The vascular system consists of the heart with a ventricle and venous sinuses. The nervous system consists of a pair of widely separated and asymmetric cerebral ganglia, a pedal ganglion formed by the fusion of paired, padal ganglia and single parieto splanchnic ganglion, from which nerves lead to different parts of the body. The excretory system is composed of a pair of asymmetric nephridia which are connected dorsally by a short, transverse channel; the right nephrium has a long caecum posteriorly while the left one has only a short caecum. The nephiridia open separately to the exterior at the ventral extremity of each renal tube near side is highly irregular mass composed of a large lobe covering the stomach, digestive gland and coil of the intestine and number of other lobes spread out in various directions in the mantle. The gonad opens in to the inter nephridial passage by a small aperture. Sexes are separate. Windowpane releases 24 million eggs at a single spawning and 250 million eggs during spawning season. The eggs diameter is given as 70 micron.

Anatomy of Window Pane Oyster

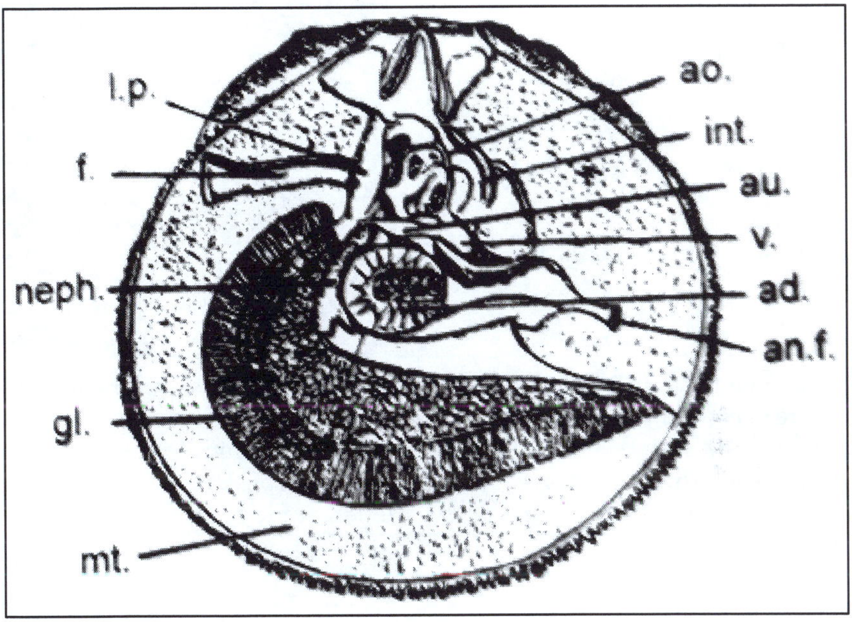

l.p.: Libial palps; An. F.: Anal funnel; f.: Foot; mt.: Mantle; int..: intestine; ad.: Adductor muscle; a.o.: Aorta; gl.: Gills; neph: Nephridia; v.: Ventricle.

Scientific Name

Poromya granulate Nyst.

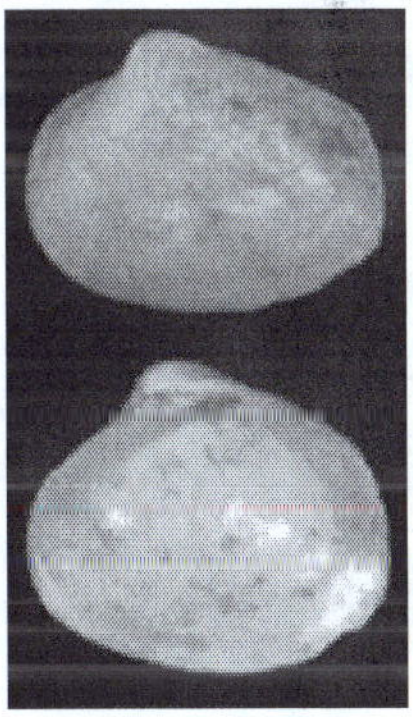

Classification

Phylum	:	Mollusca
Class	:	Bivalvia
Order	:	Anomalodesmata
Super Family	:	Poromyoidea
Family	:	Poromyacea
Genus	:	Poromya
Species	:	*granulata*

The shell is 1/4 to 1/3 inch in length, ovate, inflated and fragile. Beaks inflated and tuned forward. Exterior cream white, with and irregular coating of fine granules which resemble sugar coating. In fresh materially this granular deposit is also found on the inner margins of the valves. There is slightly gaping at the posterior end. Interior of valve is silvery white. They borrow in to mud. The specimen is carnivorous and feeds by sucking in food with an adapted muscular gill. It probably feeds mainly on dead crustaceans.

Scientific Name

Protothack staminea Conrad

Classification

Phylum	:	Mollusca
Class	:	Bivalvia
Sub-Class	:	Heterodonta
Order	:	Veneroida
Family	:	Veneridae
Genus	:	Protothack
Species	:	*staminea*

The shell is 1 ½ to 2 inches in length, sub ovate, beaks nearer the anterior end, sculpture of concentric and radial ribs which form beads as they cross each other at the anterior end of the shell. Radial ribs stronger on the middle of the valves. Beaks are almost smooth. Exterior is rusty brown with a purplish cast. It is very abundant, wide spread species with a number of varieties. Some times with a mottled dolor pattern

Scientific Name

Solen lamarckii Desh

English Name

Razor shell

Classification

Phylum	:	Mollusca
Class	:	Bivalvia
Sub-Class	:	Heterodonta
Order	:	Incertea sedis
Super Family	:	Solenoidea
Family	:	Solenidae
Genus	:	Solen
Species	:	*lamarckii*

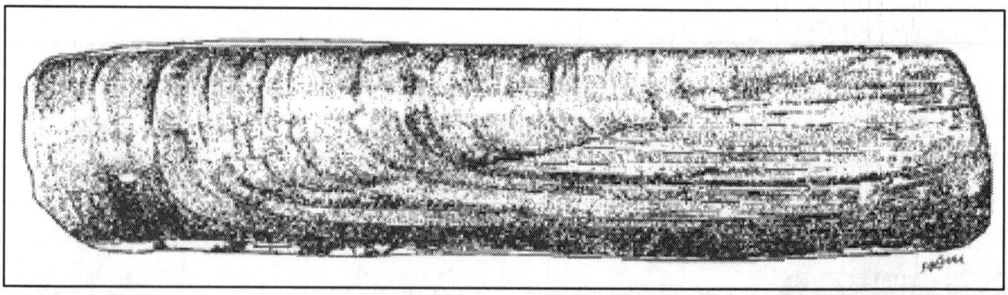

Diagnostic Characters

Shell equivalve, laterally compressed to subcylindrical in cross-section, with a narrowly elongate, often sword-like or razor-like shape, very inequilateral and gaping at both ends. Umbones not prominent, frequently near the anterior end of dorsal margin. Outside of shell is essentially with concentric growth marks, often changing abruptly in direction along a diagonal line running from the umbones to the posteroventral end of valves. Periostracum is prominent, frequently glossy. Ligament external, set in an elongate linear groove of posterodorsal margin. Hinge feeble, are with at least 1 peg-like cardinal tooth in each valve, and sometimes 1 or 2 lamellar posterior teeth. Interior of shell porcelaneous. Two dorsally placed, narrowly elongate to rounded, adductor muscle scars, the anterior one commonly larger. Pallial sinus relatively shallow, its ventral, and even sometimes dorsal limb largely confluent with pallial line. Internal margins smooth. Gills of eulamellibranchiate type, posteriorly placed, with smooth or folded branchial sheets. Foot long and narrow, often terminally swollen. Siphons are naked, short to long, fused, at least at their base. Mantle margins widely fused ventrally, with an anterior to anteroventral opening.

Scientific Name

Solen truncates Wood

English Name

Razor shell

Classification

Phylum	:	Mollusca
Class	:	Bivalvia
Sub-Class	:	Heterodonta
Order	:	Incertea sedis
Super Family	:	Solenoidea
Family	:	Solenidae
Genus	:	Solen
Species	:	Truncates

They are marine as well as brackish water shells. Shell equivalve, laterally compressed to subcylindrical in cross-section, with a narrowly elongate, often sword-like or razor-like shape, very inequilateral and gaping at both ends. Umbones not prominent, frequently near the anterior end of dorsal margin. Outside of shell is essentially with concentric growth marks, often changing abruptly in direction along a diagonal line running from the umbones to the posteroventral end of valves.

Scientific Name

Tagelus califormianus Gray 1847

Classification

Phylum	:	Mollusca
Class	:	Bivalvia

Order : Heterodonta

Super Family : Tellinoidea

Family : Solecurtidae

Genus : Tagelus

Species : *claiformianus*

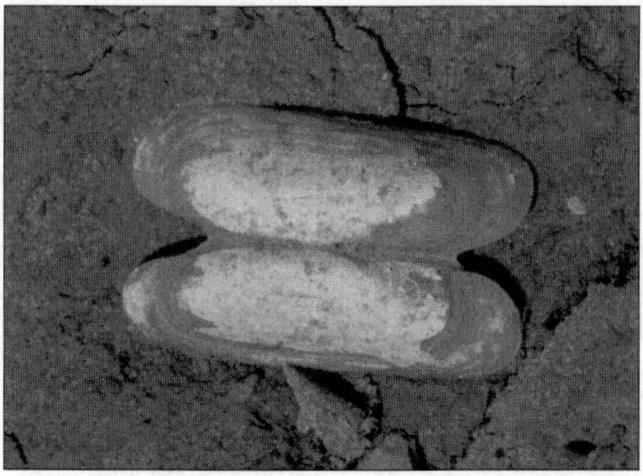

The shell is 2 to 4 inches in length. Pallial sinus does not extend past a line vertical to the beaks. External colour is yellowish white under a dark brown periostracum which is readily striated. Interior is white. Common on muddy sand flats near marshes. They live 8 to 10 inches below the surface of the sand.

Scientific Name

Tapes aureus

Classification

Phylum : Mollusca

Class : Bivalvia

Order : Heterodonta

Super Family : Veneroida

Family : Veneracea

Genus : Types

Species : *aureus*

The shell is solid, medium size 1¾ inches, moderately convex valves, roughly rhomboidal, equivalve and in equilateral. The periostracum is thin and glossy. The colour can be pale yellow, white or chest nut with broken zigzag lines, or dark chestnut patches. On the other hand, the inside of the valves in consistently golden yellow, hence its specific name. The pallial line has a large sinus, which does not reach the middle of the shell. Body has a pair of gills each of which made up of two series of lamellae, entirely fised by interlanellar junctions. The foot is normally dark, and the two siphons of moderate length, have their aperture surrounded by cirri. They are found in sandy, sandy-muddy or muddy bottom in the upper level of the intralittoral zone. They are actively fished for food.

Scientific Name

Tellina ala Gravely

Classification

Phylum : Mollusca

Class : Bivalvia

Sub Class : Heterodonta

Order	:	Veneroida
Super Family	:	Tellinacea
Family	:	Tellinidae
Sub Family	:	Tellininae
Genus	:	Tellina
Species	:	*ala*

Reference

Tellina ala Gravely, *Bull. Mad. Govt. Mus. (Nat. Hist) I* No1, 1927, P. 106

The shell is readily distinguished from that of the preceding species by its smaller size, less strongly infected valves and by the hind margin being pointed and beak-like. Except for the angular hind margin, the shell bears a very regular and evenly rounded outline. As in the preceding species the shell bears a radial keel just beside, the posterior margin, preceded sculpture is relatively less emphasized in the region of the preceding species. The umbo is nearly medial and the dorsal margin behind it slopes obliquely, while it is almost horizontal in front of the umbo. The surface of the shell is uniformly traversed by regular, very fine, close-set, almost microscopic, concentric striate which render the surface more or less glossy. The umbo is small, sharply pointed and slightly inclined posterly.

Scientific Name

Tellina meropis Dall

Classification

Phylum	:	Mollusca
Class	:	Bivalvia
Sub Class	:	Heterodonta
Order	:	Veneroida
Super Family	:	Tellinacea
Family	:	Tellinidae
Sub Family	:	Tellininae
Genus	:	Tellina
Species	:	*meropsis*

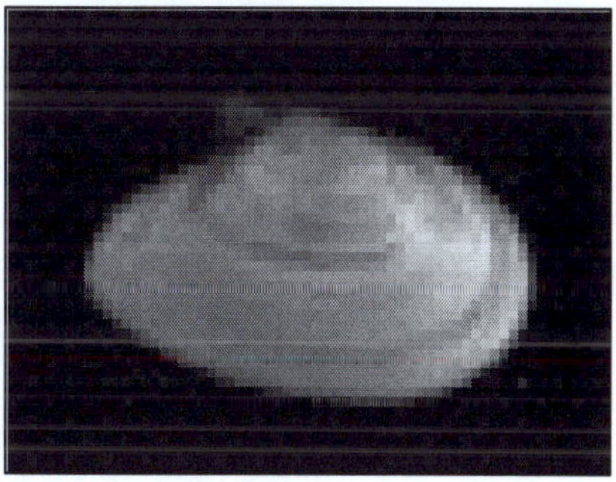

They are 1/2 inch in length, ovalish, pure white, smoothies, with exceedingly fine growth lines. Surface silky, but rarely with and iridescent shown. Beaks are slightly towards the posterior end. Ligaments are not prominent and light – brown.

Scientific Name

Tellina idae Dall

Classification

Phylum	:	Mollusca
Class	:	Bivalvia
Sub Class	:	Heterodonta
Order	:	Veneroida
Super Family	:	Tellinacea
Family	:	Tellinidae
Sub Family	:	Tellininae
Genus	:	Tellina
Species	:	*idae*

The shell is 2 to 2½ inch in length, elongate, compressed. With strong rather evenly spaced, concentric, lamellate threads posterior end narrow, slightly twisted and with a rounded, radial ridge near the dorsal margin (in right valve) or a ridge at the dorsal margin and a furrow below it (left valve). Ligament is elongate and sunk deeply into the long deep dorsal margin furrow colour grayish white.

Scientific Name

Trachycardium egmontianum Shuttleworth 1856

Classification

Phylum	:	Mollusca
Class	:	Bivalvia
Sub Class	:	Heterodonta
Order	:	Veneroida
Super Family	:	Cardiacea
Family	:	Cardiidae
Genus	:	Trachycardium
Species	:	*egmontianum*

The shell is 2 inches in height with 27 to 37 strong, prickly, radial ribs. It is externally whitish to tawny gray with odd patches of weak yellow, brown or dull purple. Interior is glossy, comely brightly hued with salmon, reddish and purple. Don't confuse with muricatum which is more oval, has more ribs which are not sharply scaled at the counter of the shell and is commonly only yellowish inside. It is common shallow water species, especially on the Gulf of Kutch.

Scientific Name

Venus verucosa

Classification

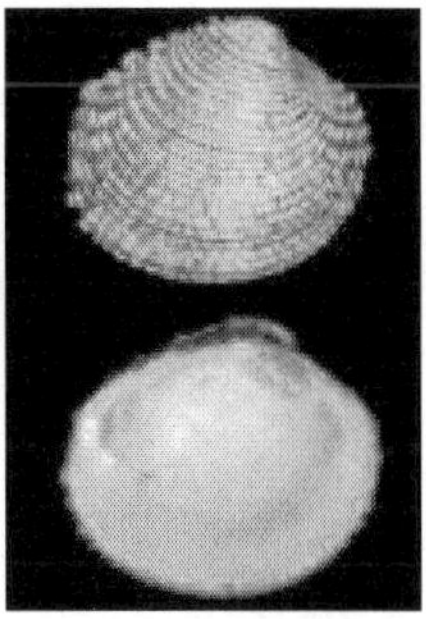

Phylum	:	Mollusca
Class	:	Bivalvia
Sub Class	:	Pleotaxodonta
Order	:	Veneroidae
Super Family	:	Veneracea
Family	:	Verneridae
Genus	:	Venus
Species	:	*verucosa*

Size

Medium size (2½ inch) round outline, equivalve and inequilateral.

The periostracum is thick, not very persistent and brown in colour. The sculpture consists of concentric lamellar ribs, which are broken up posteriorly into blunt processes. Between the ribs there are thin radial grooves that give a reticulated appearance to young examples. The ligament is external and posterior to the umbones, the hinge is typical heterodont, with three cardinal teeth on each valves.

Body

It has a pair of gills, each of which consists of two series of lamellae, extensively fused by interlanellar junctions. Normally, the foot is dark, the two sipnones are of moderate length, as the animal does not borrow deeply.

Habitat

V. verrucosa normally lives on sandy or fairly coarse detrital bottom in the infralittoral zones, but it can extend into the circulittoral zone.

Scientific Name

Venus chemnitzii Hanlay

Classification

Phylum	:	Mollusca
Class	:	Bivalvia
Sub Class	:	Paleotaxodonta
Order	:	Venvoida
Family	:	Veneridae
Sub Family	:	Venerinae
Genus	:	Venus
Species	:	*Chemnitzii*

Reference

Venus chemnitzii, Hanley, Proc. Zool. Soc. London, 1844, P. 169

The concentric laminae are thinner, eraot and closely folded on their lower surfaces, the concavity of these folds facing away from the umbo. The intera-spaces between these laminae are moderately wide, flattened and traversed by close-set, narrow radial ridges. The hinge teeth are disposed much as in *V. reticulate,* but are thinner and the middle cardinal tooth is not bifid. The pallial sinus is deep angular at the apex. The shell is whitish or very pale yellowish and rayed with three or four broad, radial bands, each consisting of a series of brownish patches, triangular towards the umbo and squared ventrally, the latter often coalescing with each other.

Scientific Name

Venus reticulate Linne

Classification

Phylum	:	Mollusca
Class	:	Bivalvia
Sub Class	:	Pleotaxodonta
Order	:	Veneroida
Super Family	:	Veneracea
Family	:	Veneridae
Sub Family	:	Venorinae
Genus	:	Vinus
Species	:	*reticulata*

Synonyms

Venus corbis, Lamarck,

Venus reticulate, Hornell

Reference

Linne, Syst. Nat., Ed. XX 1758, P. 687

The shell is thick, solid and closely reticulately sculptured throughout, with strong, close set, concentric, granulated ridges and with weaker radial ridges in the inter sties between them. The surface consequently presents a coarse, almost tuberculated appearance. The hinge is thick, and bears three stout cardinal teeth, the middle one of which is distinctly bilged. The hinge area is of a characteristic bright orange red colour. The tooth in front of the cardinala in the left valve and hallow in the right are entirely absent. The pallial sinus is broad and U-shaped. The shell is white, marked with brown spots and blotches which tend to be arranged radically. The front margin of the shell is rounded and narrower than the broad and truncated hind margin. The reticulated sculpture of the shell and the orange colour of the hing teeth are very characteristic of this species and help to distinguish even worn out shells washed upon the beach or lodged among the shingle on the reefs.

Scientific Name

Venus papilionacea, Lamark.

Classification

Phylum	:	Mollusca
Class	:	Bivalvia
Sub Class	:	Pleotaxodonta
Order	:	Veneroida
Super Family	:	Veneracea
Family	:	Veneridae
Sub Family	:	Venorinae
Genus	:	Vinus
Species	:	*papilionacea*

Synonyms

Phaphia ala- papiliones Roding

Tapes rotundata Reeva 1864.

Paphia ala-papiliones, Gravely 1941

Reference

Vinus papilionácea Lamark,, *Anim. Sans.Vert., V,* 1881, P. 594

In general appearance and proportions of the height and length, the shell of this species closely resembles that of the preceding species but the posterior margin is slightly broader in the present one. The ventral margin is almost straight and the front and hind margins are evenly rounded. The surface of the shell is glossy and traversed by regularly spaced concentric grooves, separated by flattened concentric ridges. These concentric grooves lose their distinctness towards the posterior margin, the posterior one-fourth part of the hinge teeth are as in the preceding species.

Chapter 3
Marine Gastropoda

Introduction

Molluscs are very large group of animals and they are subdivided into number of groups and class. One of the big classes is Gastropod. The gastropods are the largest and most diverse of all the classes of the molluscs, containing an enormous number of species that can live as free-floating animals hundreds of miles off the coast to intertidal, estuarine or land animals.

Classification

The gastropods are subdivided into three groups call sub-classes: *Prosobranchi, Opisthobranchia* and *Pulmonata.*

Prosobranchis

The simplest *prosobranchis* have paired gills, kidney and collecting chambers in the heart (auricles) and include the limpets, which are herbivores with simple cone-shaped shells, the top shells which are coiled-shelled herbivores. A further development in the *prosobranchis,* with the loss of the paired gills and kidneys and auricles gives the group that contains many of the most collected gastropods. In this that trend from herbivores to carnivores can discerned. They include the periwinkles, helmet shells, conches and cowries. Many have well developed sense organs such as eyes.

Opisthobranchs

The *opisthobranchs* include many beautiful molluscs many of which are shell-less. The shells from animals in this group are often delicate and do not provide any real protection for the animal. In many cases their bodies are large and gelatinous and this reduction in weight of the shell has enabled many of these creatures to swim.

Some of this group, freed from the restrictions of crawling on the bottom, have evolved a wide range of feeding patterns, including predation.

The shell-less *opisthobranchs*, the sea slugs, are a beautiful and interesting group of animals. They are successful and yet they do not possess a protective shell.

Pulmonates

The most highly evolved gastropods are the *pulmonates* which have successfully invaded the land. During the transition from marine to terrestrial life they have lost their gills and some have then gone on to invade freshwater to produce most fresh-water snails. Thus the air-breathing lung produced during the conquest of the land might then be considered a handicap to these fresh-water snails. Most land molluscs have shell and are called snails to distinguish them from the shellless slugs. Some *pulmonates* have reinvaded the sea and in warm countries many of the upper shore limpets are *pulmonates.*

Limpet

The term limpet is applied to a large group of common gastropod molluscs that have a relatively simple single shell, or not a tall coiled spire. Limpets are usually found stuck closely on to rocks on the shore or in shallow water. Although very similar in appearance, these shells may belong to unrelated groups.

The limpets can be sub divided by some very obvious shell characters – those with holes in their shells, those with no such holes and no internal ledges, the shell with internal ledges as in the following text.

Abalones

In all these limpets, which have holes in their shells, water is drawn over the gills from underneath the shell and is exhaled through the shell apertures. These animals seem to be more susceptible to desiccation than the other groups of limpets. They include abalones or ormers, which are widely prized both for eating and for their wonderfully lustrous shells. The abalones have retained the primitive two-gilled condition. They graze on algae and detritus from the surface of rocks and are not intertidal, being normally found from the extreme low water of the spring tides down to about 12 m in depth. Abalones are unisexual and reproduce in rather a primitive way by external fertilization: the gametes are shed into the water. It is thought that the release of sperms into the water triggers the shedding of the eggs.

The other group of limpets with shell perforation are the key hole limpets. In this group of animals the anus ends near the exit hole for the respiratory water current in the top of the shell. In this way the faeces are swept well clear of the animal and the inhalant water current does not become contaminated by them.

Anatomy

The anatomy of the soft parts of gastropods is similar in all species. They usually have a head with tentacles on its upper surface. Eyes can be found at the base of the tentacles and on the underside of the head is the mouth. Gastropods do not have conventional jaws but abrade their food with their rasp-like redula. Small pieces of

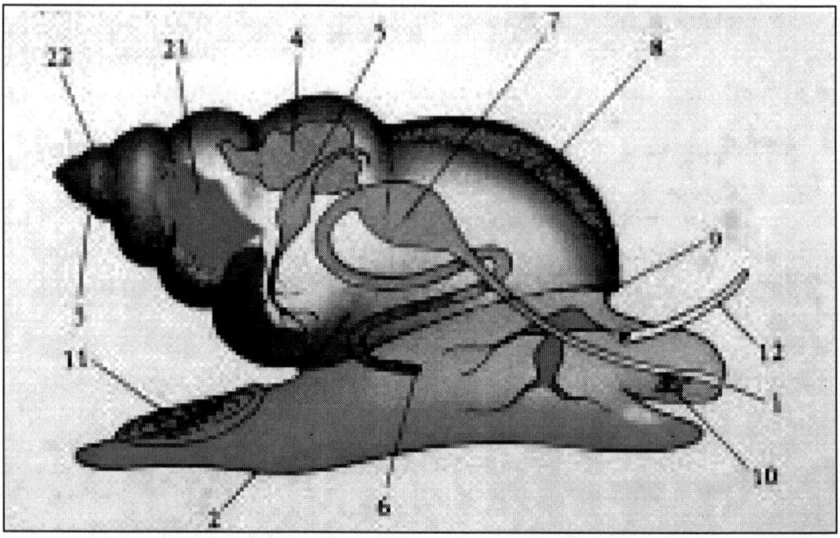

1: Mouth; 2: Foot; 3: Gonad; 4: Kidney; 5: Heart; 6: Anus; 7: Gut; 8: Gill; 9: Mantle; 10: Radula; 11: Operculum; 12: Head tentacles; 13: Feeding tentacles; 14: Retractor muscle; 15: Inhalant; 16: Exhalant siphon; 17: Inhalant muscle; 18: Palps; 19: Adductor muscle; 20: Hinge; 21: Digestive gland; 22: Shell.

food are then swallowed and digested in the gut, undigested food being excreted from the anus which opens in the mantle cavity. They have a blood vascular system with a heart, but their blood is usually blue and not red like that of vertebrates; this is because the oxygen is carried from the fills to the tissues by a copper-based blood pigment called haemocyanine. They usually have a kidney which is responsible for the excretion of excess nitrogen containing compounds, derived from the digestion of protein. The kidney duct, the ureter, also opens in the mantle cavity near the anus.

Reproductive System

The primitive archaeogastropods retain two nephridia; the right nephridium provides the passage for eggs or sperm from the ovary or testis to the mantle cavity. The sexes are separate in nearly all prosobranchs, although in a few taxa, such as *Crepidula,* an animal begins life as a male and then changes to a female later. Opisthobranch and pulmonate species are hermaphroditic and often protandrous (male gonads maturing first); however, in many taxa, adults become simultaneous hermaphrodites (male and female gonads are functional at the same time). Internal fertilization is common in the more advanced marine species.

Gastropods often have quite complicated reproductive systems, mainly because they are hermaphrodites – bisexual animal producing both eggs and sperms. The more advanced ones are unisexual and often fertilization is internal, the sperms being transferred to the females during a lengthy mating process.

Nervous System

The nervous system is simple and the head is completely absent, reflecting the sedentary habit. In primitive bivalves (*e.g.*, Palaeotaxodonta) there are four pairs of ganglia–cerebral, pleural, pedal, and visceral. In all other bivalves the cerebral and pleural ganglia are fused into two cerebropleural ganglia, located above and on either side of the esophagus. The pedal ganglia are in the base of the foot, and the visceral ganglia are located under the posterior adductor muscle. Nerve fibres arising from the cerebropleural ganglia extend to the pedal and visceral ganglia.

These animals are capable of some quite complicated responses to the environment which, especially for inter tidal species, is vary variable. The ability to move and search for food, the avoidance of predator and the reaction to changing environment have all led to the development of a sophisticated central nervous system and "brain". In gastropods this takes the form of a series of nerve ganglia arranged in a ring around the oesophagus at the front end of the gut.

Respiratory System

The respiratory system of gastropods varies greatly in form. These variations were once used as a basis for dividing the group into subclasses. The majority of marine gastropods breathe through a single gill, supplied with oxygen by a current of water through the mantle cavity. This current is U-shaped, so that it also flushes waste products away from the anus, which is located above the animal's head, and would otherwise cause a problem with fouling.

In primitive groups, such as certain limpets, water is drawn underneath the shell over the gills, which surround the muscular foot, by the pumping action of numerous hair like cilia. In others, such as the keyhole limpets and the abalones, holes are found in the shell through which the animal expels water. However, this not varies satisfactory shell modification is not a widely adopted method of gill ventilation and such species are unable to colonize the inter-tidal zone. Probably the water loss through the holes in the shell is excessive when the animals become exposed.

Locomotion

Instead of being attached to bone or ligaments the muscles work against 'fluid skeleton' formed by the blood (haemoskeleton), which thus aids locomotion.

Shell

The modern gastropod shell is much more likely to be a very subtle exoskeleton against which the soft-bodies gastropod can extend itself out from its shell by an unscrewing action, using its hydrostratic skeleton and the muscles of the mantle. The coiled shell would therefore be an essential feature in the development of the gastropod, enabling it to withdraw into its shelter when facing a predator, or when environmental conditions are adverse. it also enable the animal to extend itself even when not immersed in water to providing a series of inclined planer to operate against when extending its body.

In more advanced gastropods, such as the whelks, water is drawn into the mantle cavity through a siphon, the mobile and fleshy snout which extends forward of the shell, enabling the animal to draw fresh and oxygenated water into their mantles even when they are feeding on carrion in the immediate vicinity of which water may be foul and deoxygenated.

In this chapter some of the important systems found in their life are discussed. Scientific name, Identification, Classification, synonyms of species etc are given for each different species.

Details of each shell like name, classification, photographs Synonyms, Reference, description, etc are given below.

Scientific Name

Acmata pelta Exchscholts

Classification

Phylum	:	Mollusca
Class	:	Gastropoda
Sub Class	:	Patellogastropoda
Order	:	Acmaeidae
Family	:	Acmaea
Genus	:	Acmata
Species	:	*pelta*

Description

The shell is 1 to 1 ½ inches in maximum diameter, elliptical in outline, with a moderately high apex which is placed 1/3 to almost ½ ways back from the anterior end. With about 25 axial, weakly developed, radial ribs. Edge of the shell is slightly wavy. External colour of strong black radial, often intertwine strips on a whitish cream background. Interior is usually faint bluish whitish with or without a dark brown spot. Interior is usually faint bluish whitish cream background. Inner edges are with alternating place and cream bars. It is very common in rock dweller.

Scientific Name

Achatina fulica Bowdich.

Classification

Phylum	:	Mollusca
Class	:	Gastropoda
Order	:	Stylommatophora
Super Family	:	Helicarionidae
Family	:	Achatinidae
Genus	:	Achatina
Species	:	*fulica*

Reference

Achatina fulica Krishnachandra Ghose, *Bom. Nat. Hist. Soc.*, 1963 Vol. 60 (1): 228 – 232.

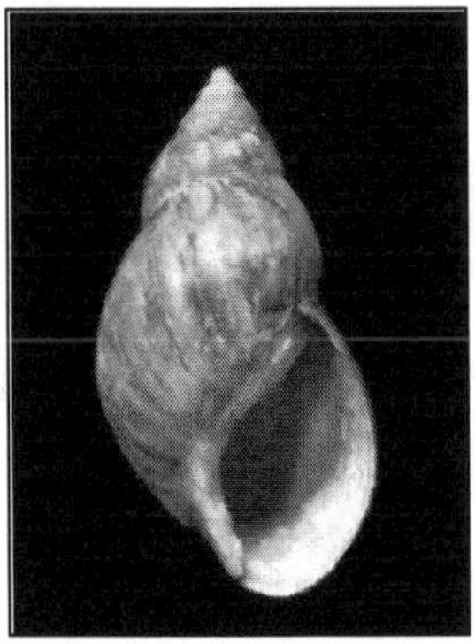

Description

The shell has a conical shape, being about twice as high as it is broad. Either clockwise (sinistral) or anti-clockwise (dextral) directions can be observed in the coiling of the shell, although the right-handed (dextral) cone is the more common. Shell colouration is highly variable, and dependent on diet. Typically, brown is the predominant colour and the shell is banded. The colouration of the shell is influenced to some extent by the climatic conditions, but sunlight seems to have no effect on pigmentation during early stages of growth. They attain sexual maturity at the age of about six months. In brood, the first young snail comes out of the egg within 1 to 14 days after laying. In most cases the young snail begin feeding on soil, leaves of plants.

The adult snails have a height of around 7 centimetres (2.8 in), and their length can reach 20 centimetres (7.9 in) or more.

Distribution

Andaman and Nicobar Islands

Areas of Abundance

Andaman and Nicobar Islands and West Bengal.

Scientific Name

Angaria plicata Kiener 1836

Classification

Phylum	:	Mollusca
Class	:	Gastropoda
Order	:	Archaeogastropoda
Family	:	Turbinidae
Sub Family	:	Angariinae
Genus	:	Angaria
Species	:	*plicata*

Description

This is moderately large, thick, somewhat massive, reddish shell with very rough, coarse looking surface. The body whorl is large and of considerable width; it is covered with blunt, spiny recesses all over but spines are much less developed than in most other species of this genus. The spire is low and the whorls are more or less sharply angular about their middle, being there by divided into a horizontal upper part which bears thick, transpiral folds and vertical lower part. The surface is traversed throughout by strong spiral ridges of which these on the lower art of the whorls are often drawn out into short spiny processes, but the later may be indistinct or even absent when the shell is worn out.

Scientific Name

Architectonica nobilis Roding

Classification

Phylum	:	Mollusca
Class	:	Gastropoda
Sub Class	:	Prosobranchia
Order	:	Mesogastropohoda
Super Family	:	Architectonicacea
Family	:	Architectonicadae
Genus	:	Architectonicea
Species	:	*nobilis*

Description

The shell is 1 to 2 inches in diameter, heavy, cream with reddish brown spot which are especially prominent just below the structure. Sculpture of 4 to 5 spiral cords which are usually beaded. Umbilicus round, deep and bordered by a heavily beaded, spiral cord. Operculum, corneous is with thin pucispiral brown and with lamellate growth lines. Commonly found in sand below low water line.

Scientific Name

Astraea semicostatum Kiener L. C. 1850

Classification

Phylum	:	Mollusca
Class	:	Gastropoda
Sub Class	:	Orthogastropoda
Order	:	Vetigastropoda
Super Family	:	Turbinoidea
Family	:	Turbinidae
Sub Family	:	Astraaeinae
Genus	:	Astraea
Species	:	*semicostatum*

Synonyms

Trochus stallatus, Philippi

Reference

Astraea semicostata, Pilsbry in *Tryon, man,cinch.*, X,1889,P.227

Description

The shell resembles very much that of a typical Trochus, being conical and trochiform with a strongly flattened base. The upper part of surface of the whorls above the spinouts row bears feebly developed, rather widely separated, oblique transpire ribs. The shell is pale yellowish brown, often bleached into a uniform dull white in empty shell washed upon the beach, which alone are represented in the collection.

Scientific Name

Bulla striata Bruguiere

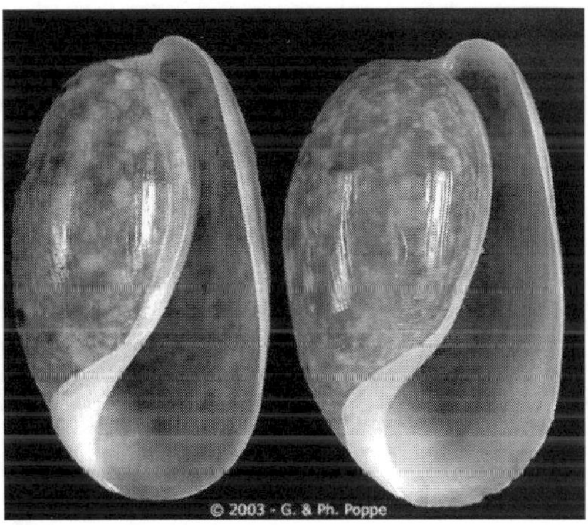

Classification

Phylum	:	Mollusca
Class	:	Gastropoda
Sub Class	:	Opithobranchia
Order	:	Cephalaspidea
Super Family	:	Cephalaspidea
Family	:	Bullidae
Genus	:	Bulla
Species	:	*striota*

Description

The species are 3/4 to 1 inch in length, similar to occidentalise, but larger, heavier and with the spiral grooves well-marked towards the base of the shell and within the apical perforation. The whorls are compressed at the apical end. Columella usually with a brown- stained callus.

Scientific Name

Bulla occidentalis A. Adams

Classification

Phylum	:	Mollusca
Class	:	Gastropoda
Sub Class	:	Opithobranchia
Order	:	Cephalaspidea
Super Family	:	Cephalaspidea
Family	:	Bullidae
Genus	:	Bulla
Species	:	*occidentalis*

Description

Shell is 1/2 to 1 inch in length, smooth, varying from fragile to quite strong and from cylindrical (young) to fairly swollen. Apex is deeply and narrowly perforated colour very variable, but usually whitish with mottling, zebra stripes and obscure bands of brown. Surface is with numerous, microscopic striations. This is a very common bubble shell which is found most easily at night and at low tide on grassy and mud flats.

Scientific Name

Bursa granularis Roading

Classification

Phylum	:	Mollusca
Class	:	Gastropoda
Sub Class	:	Oeosobranchia
Order	:	Caenogastropoda
Family	:	Bursidae
Genus	:	Bursa
Species	:	*granularis*

Description

The shell is 3/4 to 2 inch in length, flattened laterally varices axially placed one below the other. Colour orange brown with 3 narrow, white bands which appear as prominent white squares on the varices. Spiral sculpture of several rows of small beads are present, these on the periphery on the whorl having the largest beads. Teeth are aperture white.

Scientific Name

Bursa margaritula Deshayes

Classification

Phylum	:	Mollusca
Class	:	Gastropoda
Sub Class	:	Oeosobranchia
Order	:	Caenogastropoda
Sub Order	:	Tannoidea
Family	:	Bursidae
Genus	:	Bursa
Species	:	*margaritula*

Synonyms

Ranella margaritula, Landmark,

Reference

Bursa margritula, Gravely, *Bull, Mad. Govt. Hus. (Nat.Hist.)* V, No 2, 1942

Description

The shell is smaller and broader in proportion to its height than in any of the preceding species of Bursa. The varices are strongly developed, transvered by coarsely granular transverse ridges and form this continuous ridge, one on either side of the shell. The surface of the shell is ornamented throughout with finely granulates spiral ridges. The columellar border of the aperture bears a few, strong, somewhat wrinkled folds. The shell in the collection are rather faded and are uniformly pale yellowish brown, but traces of the original deep brownish colour are still seen especially on the tubercles and near the sutures.

Scientific Name

Bursa spinosa

Classification

Phylum	:	Mollusca
Class	:	Gastropoda
Sub Class	:	Orthogastropoda
Order	:	Caenogastropoda
Super Family	:	Tannoidea
Family	:	Bursidae
Genus	:	Bursa
Species	:	*spinosa*

Reference

Apte, D (1998) The book of Indian shells *Oxford University Press* 115

Description

Length of shell is 18.15-70.00 mm, diameter: 14.25-38.45 mm, max height of the aperture is 9.50-28.10. Colour of the shell is uniformly smoky brown. The shell is somewhat dorsoventrally depressed, and the whorls are slightly angulated. The most characteristic feature of the shell, however, is the presence of strong spines; each whorl has a pair of varices, one on each side of the shell as in Bursa granularis, but in the present species these varices are produced into strong, elongated, outwardly directed spines at rather irregular intervals. There is also a single spine adjacent to the posterior canal; this spine is directed backwards. In addition to the spines on the varices, there are one or two regular, spiral rows of widely spaced, short, pointed, spine like tubercles. The edge of the outer lip presents a frilled appearance owing to the presence of strong, irregular, transverse teeth.

Scientific Name

Caecum cooperi S, Smith 1860

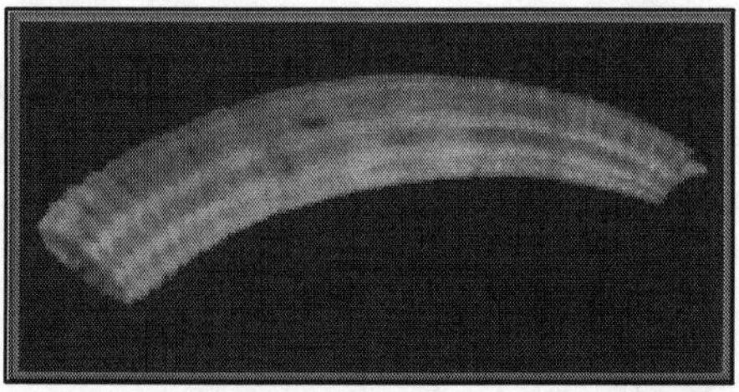

Classification

Phylum	:	Mollusca
Class	:	Gastropoda
Sub Class	:	Orthogastropoda
Order	:	Caenogastropoda
Sub Order	:	Neogastropoda
Family	:	Caecidae
Genus	:	Caecum
Species	:	*cooperi*

Description

It is 4 to 5 cm in length, slender, glossy, opaque white, with about 15 strong, longitudinal ribs. Axial raised rings are prominent near the aperture and some times give the shell a cancellate appearance at the anterior end.

Scientific Name

Cancellari reticulate Linne

Classification

Phylum	:	Mollusca
Class	:	Gastropoda
Sub Class	:	Orthogastropoda
Order	:	Caenogastropoda
Sub Order	:	Neogastropoda
Super Family	:	Cancellariidae
Family	:	Vancellariinae

Genus	:	Cancellari
Species	:	*reticulata*

Description

The shell is 1 to 1 ¾ inches in length, strong, with numerous spiral rows of small poorly shaped beads which, with the weak axial and spiral threads, give a reticulate appearance. Volumes is with 2 folds, the uppermost being very strong and furrowed by 1 or 2 smaller ridges. Colour is cream to grey with heavy, broken bands and maculations of dark orange brown. Ravels are all white. It is common in shallow water to several fathoms.

Scientific Name

Cantharus unosus Linne

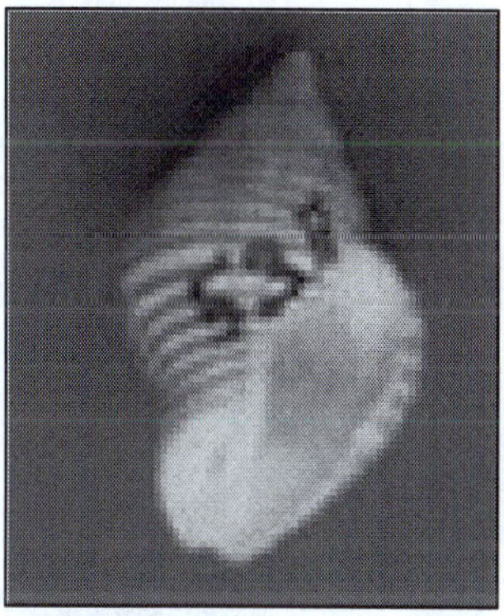

Classification

Phylum	:	Mollusca
Class	:	Gastropoda
Order	:	Neogastropoda
Family	:	Buccinidae
Sub Family	:	Pisaniinae
Genus	:	Cantharus
Species	:	*undosus*

Synonyms

Buccinum undosus Kioner

Tritonidea undosa, Schepman

Reference

Cantharus undosus, Linne, *Syst. Nat. Ed.* X, 1758, P. 740, No 409.

Description

The shell is moderately small but much larger than that of *engina.* Thick and spindle shape, strong, coloses set spiral ribs. The shell is covered when fresh, with thick periostracums which is finely but conspicuously hairy all over and which persists in the interstices between the spiral ribs. The shell is covered when fresh, with quick periostracums which are finely but conspicuously all over and which have been subjected to considerable wearing. The anterior canal is broadly open and slightly in lined. The outer lip is thickened and strongly dentate with in at the edge, the teeth merging into fine ridges interiorly. The aperture is ovate and its apex constricted into a small but well defined hotches by a ridge on either side. The presence of the periostracum renders the shell dull greenish brown, but when it is completely scraped away, the spiral ribs.

Scientific Name

Cellana radiata Born

Common Name

Limpets

Classification

Phylum	:	Mollusca
Class	:	Gastropoda
Order	:	Patellosgastropoda
Super Family	:	Nacelloidae
Family	:	Nacellidae
Genus	:	Cellana
Species	:	*radiate*

Description

They are habitat on hard rocks. Their population density varies with tidal levels. The gonads develop from February to May, and spawning begins in June and extends to February or March. Peak spawning periods are from June to August and December to February. High wind speed under optimum conditions of temperature and salinity appear to induce by MB Rao 1973 - *Cellana radiata* is immature up to 9 mm and sexes could be distinguished.

Scientific Name

Cerathium Citrium Sowely 1855

Classification

Phylum	:	Mollusca
Class	:	Gastropoda
Sub Class	:	Prosobranchia
Order	:	Mesogastropohoda
Super Family	:	Cerithiacea
Family	:	Cerithiidae
Sub Family	:	Cerithinae
Genus	:	Cerithium
Species	:	*Citrium*

Synonyms

Cerithium echinathum, Kinener, V., 1841, Cerithium, P.7.

Reference

Cerithium Citrium, Sowby, Thes. Conchyl., II, 1855, P. 179

Description

This species is considerably less available only by a few dead shells in the collection. The shell is attains a moderately large size, and has a more slender and lichgate spire than in *C, obeliscus.* The whorls are be set with strongly developed, somewhat widely separated transpiral crests and on the body whorl a more or less pronounced crests the surface bears well developed spiral ridges, and finer, thread like secondary callus on the columella thin and the anterior canal short and straight. The whorls are slightly inflated about the middle. The shell is dull brownish or grayish when fresh.

Scientific Name

Cerithium ebarneum Bruguiere

Classification

Phylum	:	Mollusca
Class	:	Gastropoda
Sub Class	:	Prosobranchia
Order	:	Mesogastropohoda
Super Family	:	Cerithiacea
Family	:	Cerithiidae
Sub Family	:	Cerithinae
Genus	:	Cerithium
Species	:	*eborneum*

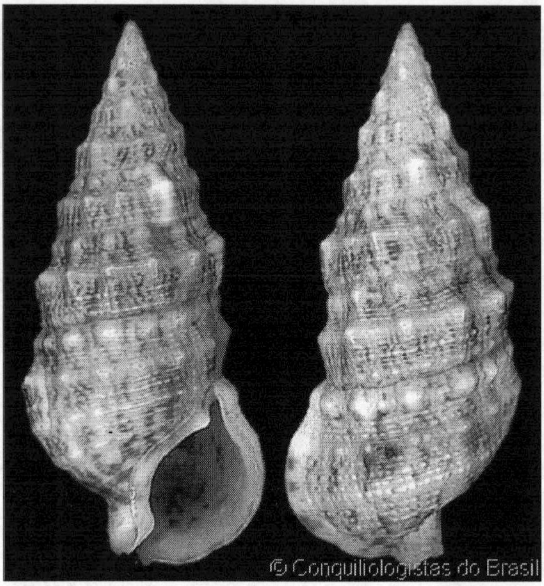

Description

The shell is 3/4 to 1 inch (rarely 1 1/2 inches) in length variable in shape, but usually moderately elongated. Each whorl has 4 to 6 spiral raws of from 18 to 22 small rounded beads. The beads are slightly larger in the middle row. There are usually a number of fairly large, former varies, colour variable, all white.

Scientific Name

Cerithium morus Landmarck 1843

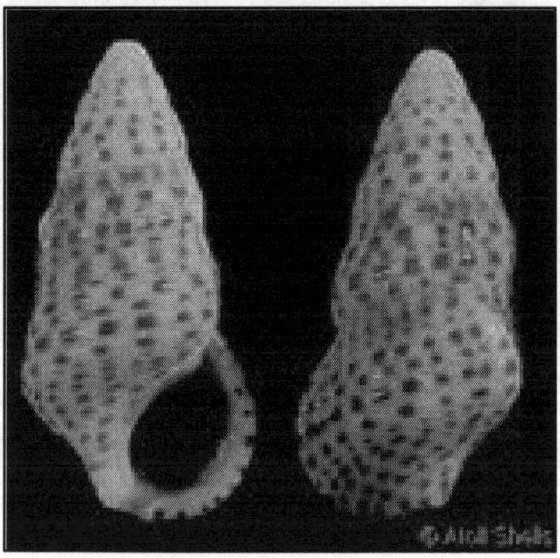

Classification

Phylum	:	Mollusca
Class	:	Gastropoda
Sub Class	:	Prosobranchia
Order	:	Mesogastropohoda
Super Family	:	Cerithiacea
Family	:	Cerithiidae
Sub Family	:	Cerithinae
Genus	:	Cerithium
Species	:	*Variable*

Reference

Cerithium Variable, Landmarck, *Aniom, Sans, Vert.(Deshayes' edit.),* 11 Vol IX, 1843, P. 302

Description

The size of the shell is 1.5 cm. The shell is rather short, ovate and presents a slightly inflected appearance. Colour is dark brown to buff, often greyish with black or brown beads which may be drawn out to form transverse lines. Each whorl of the spire is traversed by three distinct spiral rows of black, rounded tubercles, which are almost always glossy in fresh specimens. The body whorl, on the other hand, bears at least six or seven spiral rows of tubercles and sometimes also finer secondly ones in between the principal rows. The interstices between the rows of tubercles are traversed by fine, thread like, spiral strips. The aperture is short, moderately broad and more or less D-shaped. The outer lip is thick and finely ridged at the margin on the inner surface. The anterior of canal is short, but deeply excavated. The callus on the columella is distinct. The operculum is a flat oval, horny plate with a very excentric nucleus. Body whorl rounded in cross-section, not dorso-ventrally flattened, spire whorls rather straight-sided, beaded sculpture usually not axially aligned, former varix of the body whorl situated high on the left side not opposite the lips, lower part of the outer lip seems to close over the anterior canal when viewed interiorly.

Scientific Name

Cerithium obeliscus Hornell

Classification

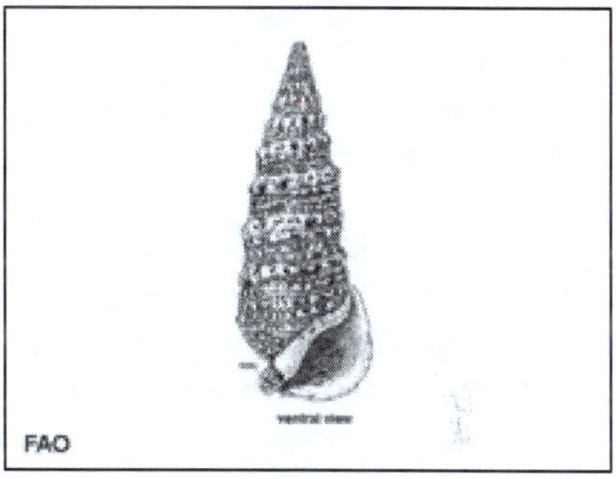

Phylum	:	Mollusca
Class	:	Gastropoda
Sub Class	:	Prosobranchia
Order	:	Mesogastropohoda
Super Family	:	Cerithiacea
Family	:	Cerithiidae
Sub Family	:	Cerithinae
Genus	:	Cerithium
Species	:	*obeliscus*

Synonyms

Vertague sinensis, Lamy, Journ. de condhyl, LXXVII, 1923

Reference

Cerithium obelisous, Hornell, *Common Molluscs of South India, Med. fish. Bull.*, XIV, 1921, P.110.

Description

The shell attains a some what large size; it is stout, with slightly arched sides, and presents rather turreted appearance. The surface of the shell is strongly sculptured with regular spiral rows of tubercles, the uppermost of these spirals being the strong and consisting of very strong, stout, raised and more or less pointed bluebells. The shell is generally pale brownish, variegated with whitish markings. This species moderately common in Gulf of Kutch, Paban, Kursadai Island.

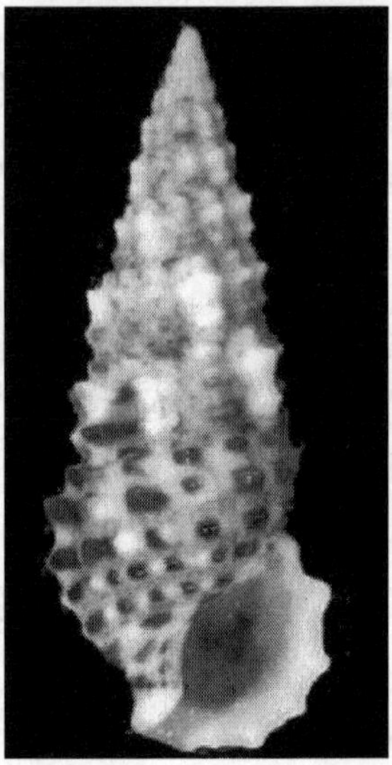

Scientific Name

Cerithium scabridum Reeve

Classification

Phylum	:	Mollusca
Class	:	Gastropoda
Sub Class	:	Prosobranchia
Order	:	Mesogastropohoda
Super Family	:	Cerithiacea
Family	:	Cerithiidae
Sub Family	:	Cerithinae
Genus	:	Cerithium
Species	:	*scabridum*

Reference

Cerithium scabridum Reeve, *Conch, Icon.*, XV, 1866, *Cerithium*, Pl. viii

Description

The shell is much smaller in all size than that of any of the preceding species and the spire is some what more slender than in *C. morus*, through not as much as elongated as in *C. citrinum*. The whorls are rather angular about the middle, but this feature is not conspicuous in some specimens, especially in young shells. The angular portion of the whorls is generally more strongly tuberoculated than the parts above and below them. The tubercles are neither quite rounded nor distinctly separated from one another, but often laterally elongated and tend to coalesce with one another forming spiral striate. The base of the shell is some what abruptly narrowed, the aperture being narrowly ovate and the anterior canal slightly produced. The shell is generally pale grayish brown, but may be occasionally darker.

Scientific Name

Cerithium variable C. B. Afdams

Classification

Phylum	:	Mollusca
Class	:	Gastropoda
Sub Class	:	Prosobranchia
Order	:	Mesogastropohoda
Super Family	:	Cerithiacea
Family	:	Cerithiidae
Sub Family	:	Cerithinae
Genus	:	Cerithium
Species	:	*Variable*

Description

The shell is 1/3 to 1/2 inch in length, not elongate. Apeartural side of body whorl some times flat. 1 to 2 for former varies found on last whorl. 3 to 4 spiral rows of even sized fine beads on the whorls of the spire. Colour dark brown black, bands of reddish brown very common under rocks in warm water.

Scientific Name

Cerithiopsis subulata Montagu

Classification

Phylum	:	Mollusca
Class	:	Gastropoda
Sub Class	:	Prosobranchia
Order	:	Neofastropoda
Super Order	:	Caenogastropoda
Super Family	:	Triphoroidea
Family	:	Cerithiosidae
Genus	:	Cerithiopsis
Species	:	*Subulata*

Synonyms

Murex subulatus, Mantagu, 1808

Description

The shell is ½ to ¾ inch in length, rather strong, slender and with about 4 – 5 whorls. Sides of whorls flattish, with 3 rows of distinct, raised, roundest beads (about 28 per row on the last whorl). There may be faint axial rib lets connecting the beads. The middle row of the beads may be reduced to a more thread in specimens from southern localities. Base is concave and with fine axial growth lines. Colour chocolate brown, with the beads a lighter shade. Some shells become eroded and coloured an ash grey or chalky brown. It is common from 1 to 33 fathoms.

Scientific Name

Columbella vulpecula Sowerby

Classification

Phylum	.	Mollusca
Class	:	Gastropoda
Order	·	Neogastropoda
Family	:	Collumbllidae
Genus	:	Collumbella
Species	:	*vulpecula*

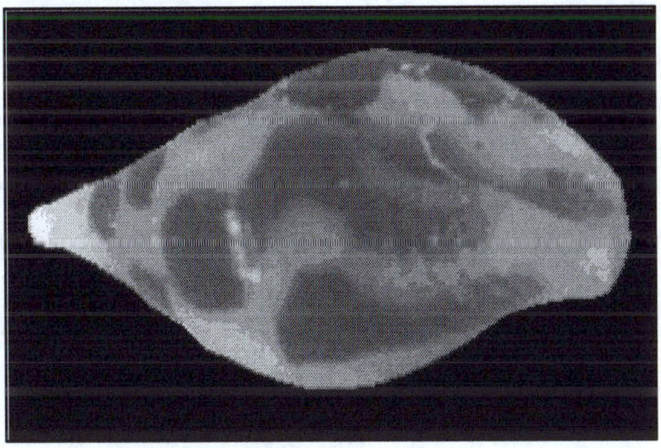

Synonyms

Pyrene vulpecula, Gravely

Reference

Collumbella vulpecula, Sowerby, *Thes. Conchv.*, I 1847, P. 1–38.

Description

Shell considerably larger and proportionately broader, with more convexly inflated whorls which are evenly rounded and not angled above, upper termination of outer lip rounded and closely approximated to and almost on a level with the general profile of the spire, outline of the outer else of outer lip more or less convexly rounded, aperture much narrowed above, shell whitish, with large, orange brown markings which coalesce to form rounded islets of the ground colour, these islets being larger and fewer on the whorls of the spire and upper half of the body whorl, and smaller, more numerous and close set in the lower part of the body whorl.

Conus

So called from the conical shape of the shell; the spire, either perfectly flat, or but slightly salient, forms the base of the cone. Its point is at the opposite extremity the aperture is narrow, rectilinear or nearly so extending from one end to the other.

Details of species are as under.

Scientific Name

Conus figulinus

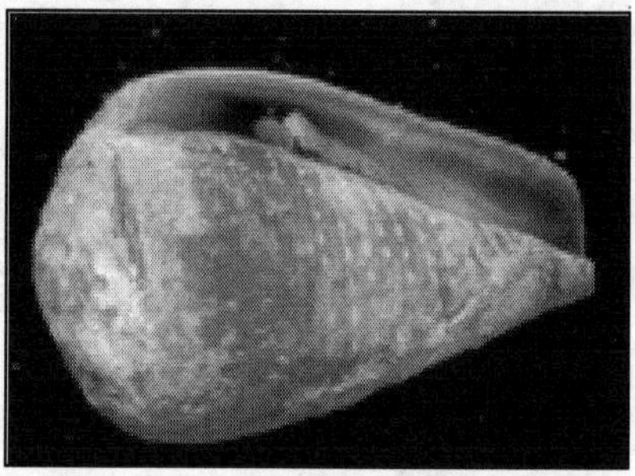

Classification

Phylum	:	Mollusca
Class	:	Gastropoda
Order	:	Neogastropoda
Super Family	:	Cinacea
Family	:	Conidae
Genus	:	Cinus
Species	:	*figulinus*

Description

Cones (Conidae), as their name implies are more or less conical in shape. The spire is usually sort and may even be telescoped so greatly as to give the apical end a truncated so greatly as to give the apical end a truncated appearance. The aperture is long and narrow, the outer lip thin, a claw-shaped small operculum. Internally the partitions are partly absorbed. The family is very large, mainly tropical in character, it includes mainly most brilliantly coloured shells, some so rare and beautiful that collectors have paid as much as Rs. 750 for single shell. Some species are dangerous to handle on current of the poisonous nature of their white but none of the Indian species have ever attempted to bite me when handling them, and none of our fishermen appear to fear them, the south sea islands on the country have a most lively dread of them, and allege that their bite is often fatal.

Scientific Name

Conus vervucosus

Classification

Phylum	:	Mollusca
Class	:	Gastropoda
Sub Class	:	Prosobranchia
Order	:	Neogastropoda
Super Family	:	Conacea
Family	:	Conidae
Genus	:	Conos
Species	:	*vervucosus*

Description

The size of this shell is 3/4 to 1 inch in length. A heavy, small cone is with a rather high spire and slightly rounded sides. It has small knobs on the shoulder of the last whorl and about 10 spiral rows of distinct warts on the sides. Colour white to yellowish with larger brown or yellow mottling.

Scientific Name

Conus araneousus

Classification

Phylum	:	Mollusca
Class	:	Gastropoda
Order	:	Neogastropoda
Super Family	:	Cinacea
Family	:	Conidae
Genus	:	Cinus
Species	:	*araneousus*

Description

This species is also found in Gulf of Kutch and they are commonly found in Ceylon. Hence numerous shells are frequently obtained to be sold as ornamental paper weights. The shell is large very thick, solid and heavy, with a broadly conical, straight sided body whorls and a more or less feebly elevated spire with concavely depressed whorls and blunt apex. The body whorl is somewhat sharply angulated towards its upper end, the portion of the whorl above this angulations being almost horizontal or even concave, and almost on a level with the lower whorls of the spire. This angular shoulder and the angular raised bottom of each whorls of the spire are coroneted, being beset with rose but these are often rubbed down in worn shells. The surface of the shell is smooth throughout except for the growth striate and a slight spiral grooving towards the base, and is even slightly glossy. The shell bears a characteristic colour pattern, it is whitish, tinged with a pale fleshy pink, and marked throughout very fine, hair like brownish lines which reticulate closely, but rather irregularly, the reticulation being generally disposed in longitudinal bands of alternately close and more open network, in addition to this, there are two or three broad, interrupted blackish brown spiral bands on the body whorl.

Scientific Name

Cymatium rebeculum

Classification

Phylum : Mollusca

Class : Gastropoda

Order : Neogastropoda

Family : Cymatiidae

Genus : Cymatium

Species : *rebecumlum*

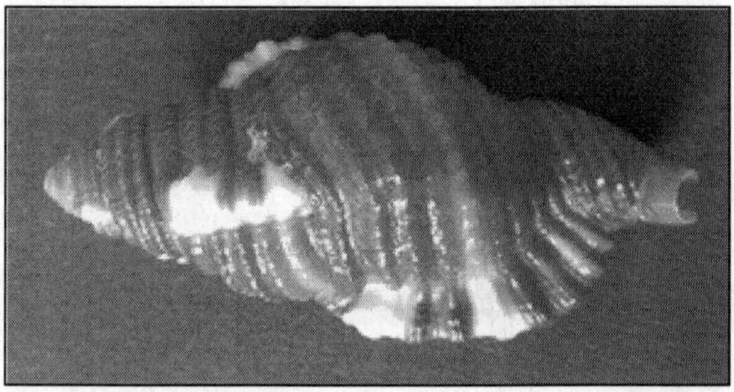

Description

The shell is small rather thick, robust, hinge spire, spindle shaped, swollen whorls and very incised suture, the aperture is ovoidal the outer lip is thickened by strong varix and has well defined and widely spaced teeth inside. The collumellar lip has a moderately thick callus with raised folds, which give way to thin teeth. The sculpture consists of spiral ribs that are as broad as the gapes between them and intersect thinner axial bands, well developed nodules are formed at the intersections of these two sculptures. The operculum is rounded and horny. Body has a rather developed interiorly truncated foot. The distinct head has a long proboscis and two tentacles with an eye at the base of each. In males the mantle cavity contains behind the right tentacles, a large penis. It lives hidden beneath or among the corals of the mid and inner reef sections of the intra littoral zone throughout the Indo-pacific province.

Scientific Name

Cypraea annulus Linne

Classification

Phylum : Mollusca

Class : Gastropoda

Order : Neogastropoda

Super Family : Cypraedcea

Family : Cypraeidae

Sub Family : Cyraeinae

Genus : Cypraea

Species : *annulus*

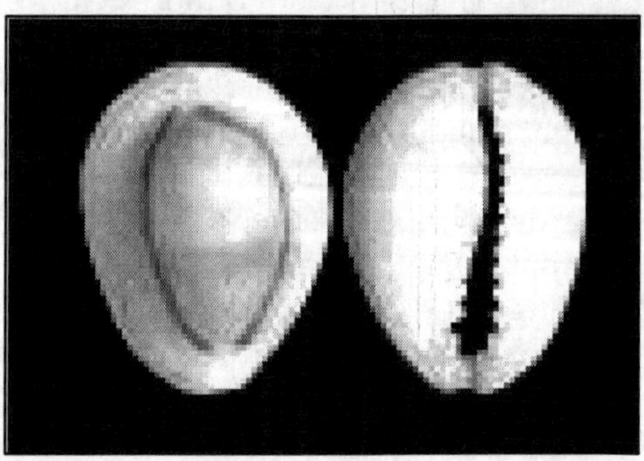

Synonyms

Monetaria annulus, Schilder, Proc. Malac. Soc., XXIII, 1938, P. 41

Reference

Cypraea annulus, Linne, *Syst. Nat., Ed.* X, 1758, P. 723.

Description

The shell is small with a slightly humped dorsum, a concavely depressed base and a more or less evenly rounded, ovate profile. The sides are thickened, whitish and bear deposit of callus, but the merge imperceptibly into the central portion of the dorsum which is of a apple bluish grey colour, encircled by an oval, yellowish ring. The columellar teeth are fine and close set while the labial ones are stronger and more markedly produced of the four races into which this species has been split up.

Scientific Name

Cypraea caurica Gravely

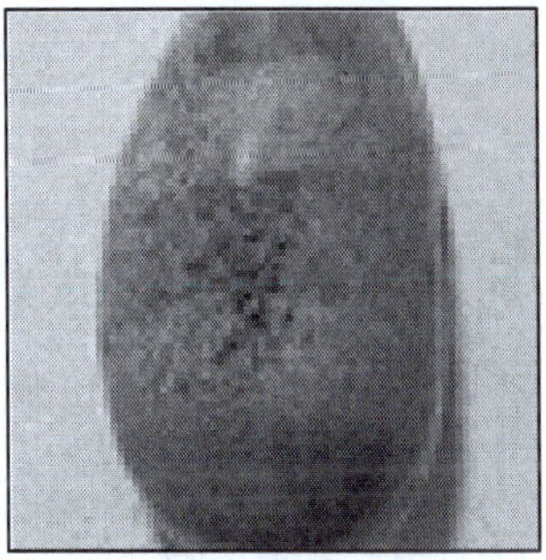

Classification

Phylum	:	Mollusca
Class	:	Gastropoda
Order	:	Neogastropoda
Super Family	:	Cypraedcea
Family	:	Cypraeidae
Sub Family	:	Cyraeinae
Genus	:	Cypraea
Species	:	*caurica*

Synonyms

Erronea caurica, Schilder, Proc. Malac. Doc. XXIII, 1938, P. 153.

Reference

Cyparea caurica, Grevely, *Bull. Mad. Govt. Mus. (Nat.Hist.), V.*no2, 1942, p.41.

Description

The shell is larger than that of *C.errones*, and usually elongated cylindrically craven with a strongly thickened with callus deposit. The sides are strongly thickened with callus deposits. The right is side being particularly strongly marginated. The dorsal surface is whitish very closely and irregularly freckled with brownish markings of which those at the centre are usually darker and coals to form a more or less diffuse and irregularly broken blotch. The sides which near a well marked callus deposit are whitish, marked with large, clouded, dark brown spots, front. The labial teeth are very stout and produced strongly cross the base of the lip, the columellar teeth are thinner, narrower closer stand much produced their outer ends being dilated.

Scientific Name

Cyparea occellata, Grevely

Classification

Phylum	:	Mollusca
Class	:	Gastropoda
Order	:	Neogastropoda
Super Family	:	Cypraedcea
Family	:	Cypraeidae
Sub Family	:	Cyraeinae
Genus	:	Cypraea
Species	:	*occellata*

Synonyms

Conus occellata, Reeva, Conch. Icon.., III, 1845, Pl. XV

Erosaria occellata, Proc. Malac. Soc. XXLII, Pt. III 1038, P-138.

Reference

Cypraea occellata, Grevely, *Bull. Mad. Govt. Mas. (Wat. Hist.)V*, No 2,1942, P.40.

Description

This is tolerably commonly species occurring at Krusadai, readily distinguished by its characteristic and beautiful pattern of colouration. The shell is roughly about the same size at that of *C. erosa*, but slightly more broadly ovate. The sides of the shell are thickened with callus and slightly reflected up, and the upper edge of the thickened margin is feebly pitted, especially on the sides of the outer lip. The dorsal surface of the shell is pale polish brown or a bright orange – yellow, ornamented with numerous rounded white spots some of which have a well marked central reddish brown spot, resembling somewhat the pupil of eye. The sides and base of shell are covered with callus, and are whitish and finally spotted with brown. The teeth are produced partly across the base of the lips 26 mm.

Scientific Name

Cypraea onyx Linne

Classification

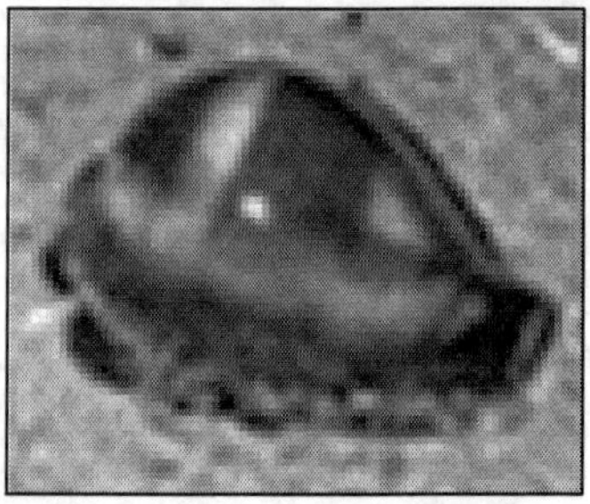

Phylum	:	Mollusca
Class	:	Gastropoda
Order	:	Neogastropoda
Super Family	:	Cypraedcea
Family	:	Cypraeidae
Sub Family	:	Cyraeinae
Genus	:	Cypraea
Species	:	*caurica*

Synonyms

Erronea onyx, Schilder, Proc. Malc. Soc. XXIII, Pt. iii, 1938, P. 149.

Reference

Cypraea onyx, Linne, *Syst. Nat. Ed. XIII*, 1791, P. 1177

Description

The shell is moderately large, thick, pear-shaped and bears exquisitely fine glosses on the surface. It is dark chestnut brown throughout, with narrow, widely separated, the sides and base are deeper chestnut, while the teeth are bright orange-yellow. The teeth are confined to the apertural margin and the aperture is dilated interiorly. Three out of the fount shell in the collection belong to this variety. The second variety, differing from the first only in its colouration is represented by a single dead shell, in this the dorsal surface of the shell's pale brownish, with two whitish, widely separated transverse bund, while the sides, base and extremities are of as much as darker brown colour.

Scientific Name

Cyprea pallid Gray 1837

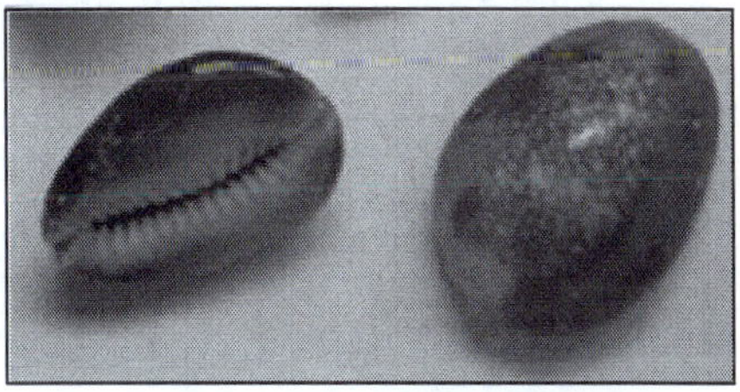

Classification

Phylum	:	Mollusca
Class	:	Gastropoda
Order	:	Neogastropoda
Super Family	:	Cypraedcea
Family	:	Cypraeidae
Sub Family	:	Cyraeinae
Genus	:	Cypraea
Species	:	*pallid*

Synonyms

Erronea pallid, Reeve, 1845

Reference

Cyprea pallid Gray, *Bull. Med. Govt. Mus.(Nat.Hist.)*,V, 1942, P. 42

Description

The shell is broadly ovate in outline and more or less pear-shape. It may be readily distinguished from *C. errones*, which is tends to resemble in colouration, by the shell being broader, shorter and more elevated and by the labial teeth of the shell being numerous and close set. The dorsal surface is pale greenish grey very minutely freckled with pale brownish marking throughout, but the centre is usually heavily mottled with dark brown, while the base is uniformly white. In the young shell there are two broad pale transverse bands across the dorsum and the aperture is dilated interiorly. The teeth are short, numerous and close-set 25 mm.

Scientific Name

Cypraea tigris Gravely

Classification

Phylum	:	Mollusca
Class	:	Gastropoda
Order	:	Neogastropoda
Super Family	:	Cypraedcea
Family	:	Cypraeidae
Sub Family	:	Cyraeinae
Genus	:	Cypraea
Species	:	*caurica*

Reference

Cypraea tigris Gravely, *Bull. Mad. Govt. Mus. (Nat. Hist.)*, V No2 1942, P. 40

Description

This specimen is represented the gallery collection by a single large dead shell. The living specimen thick, pear shaped with a strongly elevated dorsum and a slightly concavely depressed basal surface, the sides are evenly rounded off, and not much thickened with callus. The dorsum and sides are pale yellowish white, profusely ornamented with large rounded blackish pure white. The teeth are strong, the labial ones being especially stout. The aperture is referred to above is a particularly large one, measuring about 100 mm in length, but adult shell of much smaller dimensions have been collected from the Indian waters. It is obscurely banded with rows of violet-brown spots.

Large size of this specimen is frequently used as ornamental paper weight on account of their fine gloss and considerable weight.

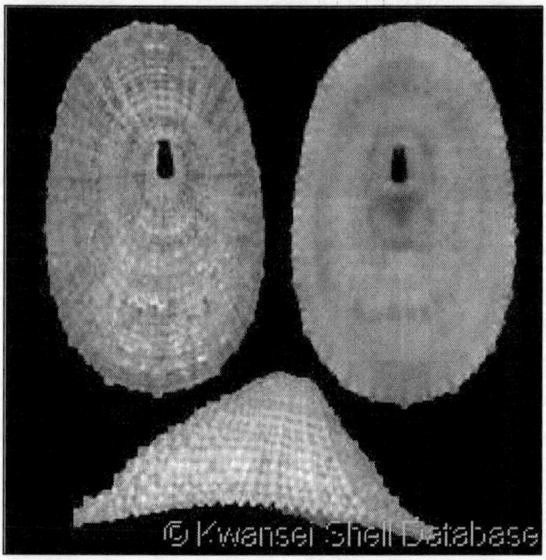

Scientific Name

Diodora dysoni Reeve, L. A.

Classification

Phylum	:	Mollusca
Class	:	Gastropoda
Sub Class	:	Orthogastropoda
Order	:	Vetigastropoda
Super Family	:	Fissurelloidea
Family	:	Fissurellidae
Sub Family	:	Emarginulinae
Genus	:	Diodora
Species	:	*dysoni*

Description

The shell is ½ to ¾ inch in maximum diameter, depressed and with straight sides. Base is ovate. Apex is slightly in front of the middle and characterized by a blunt knob situated behind the posterior wall of the small, almost triangular orifice. Sculpture of 18 strong ribs with three smaller ones between, and with numerous concentric lamellae. Colour milky white or cream with 8 solid, broken or dotted black rays. Margin sharply crenulated with the dentils arranged in 9 groups of four.

Scientific Name

Drupa modulosa C.B. Adams

Classification

Phylum	:	Mollusca
Class	:	Gastropoda
Sub Class	:	Prosobranchia
Order	:	Neogastropoda
Super Family	:	Muricacea
Family	:	Drupidae
Sub Family	:	Drupinae
Genus	:	Drupa
Species	:	*modulosa*

Description

The shell is 1/2 to 1 inch in length elongate, grossly studied with round black beads. Aperture is purplish black. Outer lip thick, and with 4 to 5 relatively large, white beads. A common shallow water species found under rocks.

Scientific Name

Litterina undulate Gray

Classification

Phylum	:	Mollusca
Class	:	Gastropoda
Sub Class	:	Orthogastropoda
Order	:	Caenogastropoda
Super Family	:	Littorinoidea
Family	:	Littorinidae
Genus	:	Litterina
Species	:	*undulate*

Reference

Litterina undulate, Gray, Zool., bookhey's voyage, 1839, P.140

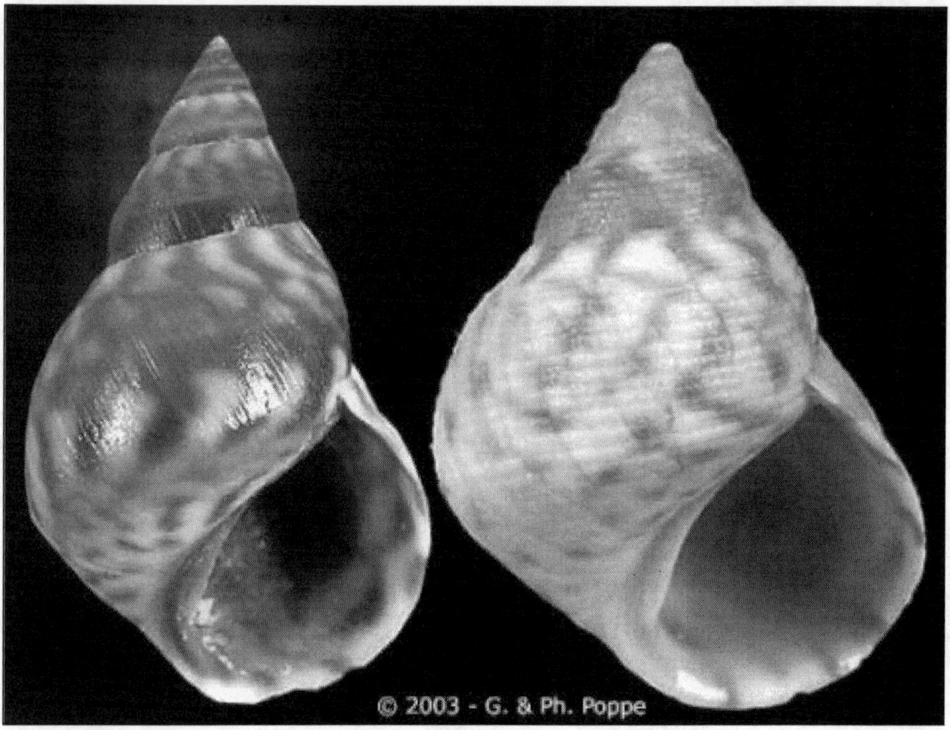

Description

The shell is generally smaller than that of the full grown *Litterina scabra* and the body whorl and base of spire are relatively narrow, the shell presents considerable variation in sculpture and colouration. The surface is sometime smooth, but more often brown in colour, but occasionally pale coloured shells, may also show fine spiral grooving zigzag, transpirally extending brownish patches.

Scientific Name

Mauriritica (Arabica) eglantine Ducles, 1833

Classification

Phylum	:	Mollusca
Class	:	Gastropoda
Sub Class	:	Prosobranchia
Order	:	Mesogastropoda
Super Family	:	Cyoraeacea
Family	:	Cypreaidae
Sub Family	:	Cypraeinae
Genus	:	Maurita
Species	:	*elanitna*

Description

The shell is medium size *i.e.* 1½ to 2 inches, thick and robust, moderately high spire. The body whorl does not completely covered the preceding ones and leaves the apical whorls exposed. The back is regularly convex, the margins are moderately so. It is this characteristic that allows this species to be distinguished easily from the rest of genus. The foot is a very dark gray siphon has its distal end extensively fringed. Commonly found under stones and corals in shallowest infra-littoral waters. It is found in Indo-Pacific area. It is found in Gulf of Kutch also.

Murex

Murex is a genus of medium to large sized predatory tropical sea snails. These are carnivorous marine gastropod molluscs in the family *Muricidae,* the murexes or rock snails.

Most Murex species live in the inter-tidal or shallow sub-tidal zone, among rocks and corals.

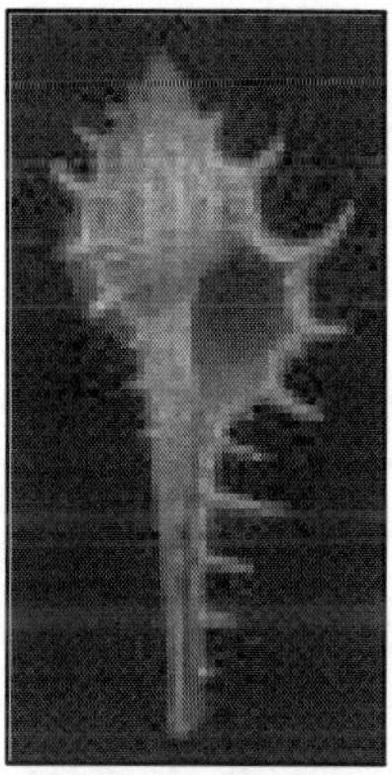

Scientific Name

Murex adustus

Classification

Phylum	:	Mollusca
Class	:	Gastropoda
Family	:	Muricidae
Genus	:	Murex
Species	:	*adustus*

Description

This species is said to be widely distributed in the Eastern area and frequently sold as an ornamental shell at Rameshwaram. The shell is readily distinguished from these of the preceding species by coal clack colour of the surface layer and by the intensely fallacious and thick – set processes with which the varies are finely and closely ridges, the interstices between which are finely and closely spirally striated. The entire surface of the shell is jet-black, except for a few traces of the brownish, almost entirely obscured ground colour which shows out here and there, especially in between the spiral ridges at the sides of the vertices. The interior of the aperture is bluish white.

Scientific Name

Murex tribulus Var. *trapa*, Gravely

Classification

Phylum	:	Mollusca
Class	:	Gastropoda
Sub Class	:	Prosobranchia
Order	:	Neogastropoda

Super Family	:	Muricacea
Family	:	Muricidae
Sub Family	:	Murianae
Genus	:	Murex
Species	:	*tríbulus Var trapa*

Synonyms

Murex trapa, Rodino, Mus. Bolten. 1798, P. 145

Murex martinianus, Reeve, Conch. Icon. III. 1845, Murex, PL XVIII.

Murex rarispina, Sowerby, Thes. Conchal. IX, 1899, Murex, P. 2

Reference

Murex Tríbulus Var. *Trapa,* Gravely, B*ull. Mad. Govt. Mus. (Nat. Hist.), V* 1942, P.49

Description

The shell is moderately thin, with a rounded, some what inflated body whorl, a well developed, turreted spire and a greatly elongated, narrow, attenuated anterior canal which is almost entirely covered up along the apertural surface is strongly spirally ridged, the ridges being slightly raised into imperfectly developed nodular swellings which are specially pronounced armed with a few, strong, moderately elongated and some what widely spaced spines, the upper ones of which generally tend to be slightly upwardly curved, the spines are finely grooved on their flattened surfaces. The outer surface of the shell is pale yellowish brown, while the interior of the aperture is pale whitish, heavily stained with dark brown.

Scientific Name

Murex virgineus Gravely

Classification

Phylum	:	Mollusca
Class	:	Gastropoda
Sub Class	:	Prosobranchia
Order	:	Neogastropoda
Super Family	:	Muricacea
Family	:	Muricidae
Sub Family	:	Murianae
Genus	:	Murex
Species	:	*virgineus*

Synonyms

Purpura virgineus, Roding, Nus. Bolten., 1798, P.141

Murex anguliferus, Lamarck, anim. Sand. Vert., IX, 1843, P. 588

Reference

Murex virgineus, Gravely, *Bull. Mad. Govt. Mus. (Nat. Hist.)*, V, No.2 1942, P.49

Description

The shell is moderately thin, with the body whorl broad above and rather strongly narrowed and below, its profiles thus assuring roughly the shape of an inverted triangle. The surface is strongly spirally ridged, the ridges being finely granular. There are three variances, some what thickened and ornamented with short curved process. Which are fairly numerous and not very widely separated. There is a large raised, Tran spirally elongated. Known like tubercle in each of the three wide interspaces between varies. The whorls are not as distinctly as angled practically closed. The surface is pale brown, while the interior of the aperture is white, tinged with pink.

Scientific Name

Nerita albicilla Linnaeus, 1758

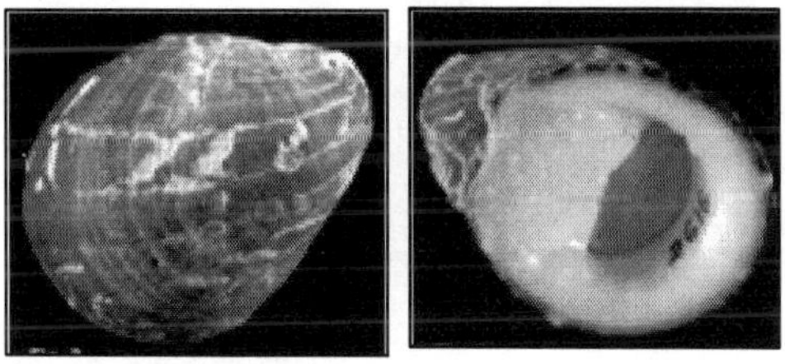

Classification

Phylum	:	Mollusca
Class	:	Gastropoda
Sub Class	:	Prosobranchia
Order	:	Neritopsina
Family	:	Neritidae
Genus	:	Nerita
Species	:	*albicilla*

Description

Shell with spire depressed, width greater than height. Sculptured are with broad, low, spiral ribs. Columellar deck pustulose, particularly at outer edge; columella concave centrally, smooth or with two or three weak teeth centrally. Outer lip is with one or two strong teeth posteriorly, sometimes one anteriorly, and 12-15 weak lira. Exterior is dull, black with white flecks or patches, often formed into two irregular spiral bands. Columella is deck white or yellow, aperture white. Operculum is calcareous, finely pustulose, grey or yellowish.

Size

Up to 24 mm in height, commonly about 16 mm.

Habitat

Shelters under rocks on open coasts between mid and low tide. Abundant in the tropics.

Scientific Name

Nerita atramentosa Reeve, 1855

Classification

Phylum	:	Mollusca
Class	:	Gastropoda

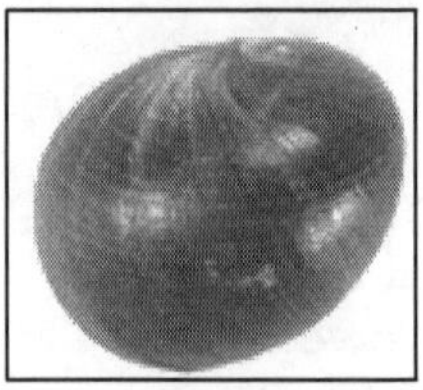

Sub Class : Prosobranchia
Order : Neritopsina
Family : Neritidae
Genus : Nerita
Species : *atramentosa*

Description

Shell with spire depressed, width equal to height. Sculptured with numerous weak and spiral grooves are often obscured by erosion. Columellar deck smooth or with weak irregular lirae; columella concave, with 2-4 elongate denticles at the centre. Outer lip is smooth or finely lirate, sometimes with a large nodule posteriorly. Exterior usually dull, uniformly black, columellar is deck white. Operculum is calcareous, tan, finely pustulose.

Size

Up to 31 mm in height, commonly about 20 mm.

Scientific Name

Natica chimniotzii

Classification

Phylum : Mollusca
Class : Gastropoda
Sub Class : Prosobranchia
Order : Mesogastropoda
Super Family : Naticacea
Family : Naticidae
Sub Family : Naticinae
Genus : Natica
Species : *chimniotzii*

Description

Depressed-ovate, spire low, umbilicus deep, with a small spiral funicle, both almost closed by a deeply grooved, thick, tongue-like reflection of the parietal callus at the posterior end of the columella. Size: Length 10.48-58.2 mm; diameter 11.45-50.3 mm. Color: Blue-grey or fawn, with a narrow orange subsutural band and sometimes faint spiral lines around the body whorl, parietal wall and callus brown, umbilicus and base white, aperture brown.

Scientific Name

Natica didya Gravely

Classification

Phylum : Mollusca
Class : Gastropoda

Sub Class	:	Prosobranchia
Order	:	Mesogastropoda
Super Family	:	Naticacea
Family	:	Naticidae
Sub Family	:	Naticinae
Genus	:	Natica
Species	:	*didya*

Synonyms

Albula didyma, Roading, Mus Biolten, 1798, P.20

Natica chemnitzii, Reclus, Ms., chenu. Illus. Conch. 1843, Pl iii

Reference

Natica didya, Grevely, *Bill. Mad. Govt. Mus. (Nat. Hist.), V*, no. 2 1942, p. 37

Description

The shell is some what large and thick, with a smooth and glossy surface, but devoid of any conspicuous colour markings. The surface is finely transpirally striated throughout. The spire is very small in proportion to the large, inflated body whorl, being almost completely depressed. The columellar border bears a large, transversely expanded, curved callosity, divided by a transverse groove into a large posterior portion which is fused through out its length to the body whorl and thus closing the umbilicus partly, and a much smaller anterior portion with a free edge, the outer surface of the shell is more or less uniformly pale brownish or grayish, while the callosity and interior of aperture are deep chestnut.

Scientific Name

Natica didyma (Roding) 1798

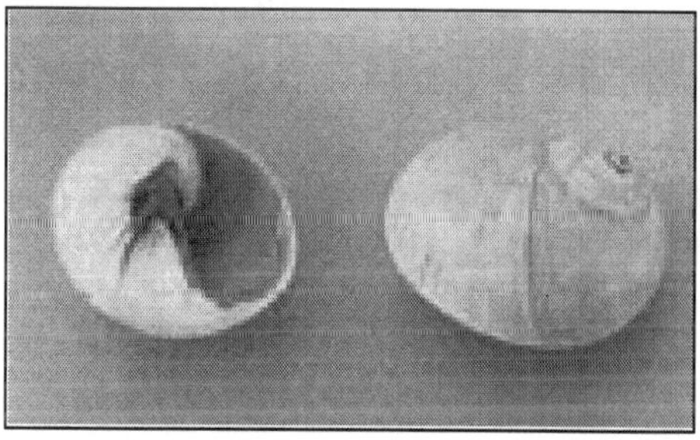

Classification

Phylum	:	Mollusca
Class	:	Gastropoda
Sub Class	:	Prosobranchia
Order	:	Mesogastropoda
Super Family	:	Naticacea
Family	:	Naticidae
Sub Family	:	Naticinae
Genus	:	Natica
Species	:	*dydyma*

Description

The shell is somewhat large and thick, with a smooth and glossy surface, but devoid of any conspicuous color markings. The surface is finely transpirally striated throughout. The spire is very small in proportion to the large, inflated body whorl, being almost completely depressed. The columellar border bears a large, transversely expanded, curved callosity, divided by a tranverse groove into a large posterior position which is fused throughout its length to the body whorl and thus closing the umbilicus partly, and a much smaller anterior position with a free edge. Color: The outer surface of the shell is more or less uniformly pale brownish or greet while the callosity and interior of aperture are deep chestnut.

Scientific Name

Nerita dombeyi Rechiz

Classification

Phylum	:	Mollusca
Class	:	Gastropoda
Sub Class	:	Prosobranchia
Order	:	Archaeogastropoda
Super Family	:	Neritidae
Family	:	Neritidae
Sub Family	:	Neritinae
Genus	:	Nerita
Species	:	*dombeyi*

Reference

Nerita dombeyi Rechiz, *Reev. Zool. Soc.*. Curiverienus, 1841, P. 149

Description

It is moderately small, with a depressed spire and obliquely enlarged body whorl, the upper part of which is more or less flattened. The surface is transverse by narrow, close-set spiral ridges, some of which, especially in the upper part of the shell, are weaker and alternate with the stronger ones. The entire lower or basal surface of the shell (*i.e.* the collumellar area and the marginal surface of the outer lip of aperture) is smooth and highly polished. The teeth at the inner margin of the columellar surface are very feebly developed. The outer lip is not toothed with in. The shell is more or less uniformly dark bluish gray, but the interstices between the ridges and paler.

Scientific Name

Narita fulgurans Gmelin

Classification

Phylum	:	Mollusca
Class	:	Gastropoda
Sub Class	:	Prosobranchia
Order	:	Neritopsina
Family	:	Neritidae
Genus	:	Nerita
Species	:	*fulgurans*

Description

The shell is ¾ to 1 inch in length, very similar to *N. tessellate* but with a lighter coloured, yellowish gray operculum. The spiral ridges on the shell are more numerous, the colour patterns blurred, the aperture relatively wider, and the teeth more prominent. This is a salt to brackish water inhabitant of protected shores, and is abundant only in certain restricted localities. It is seldom represented or properly labelled in private collections.

Scientific Name

Nerita maura Rechuz

Classification

Phylum	:	Mollusca
Class	:	Gastropoda
Sub Class	:	Prosobranchia
Order	:	Archaeogastropoda
Super Family	:	Neritidae
Family	:	Neritidae
Sub Family	:	Neritinae
Genus	:	Nerita
Species	:	*maura*

Reference

Nerita maura Rechuz, *Rev. Zool. Soc. Cuvierienne*, 1841, P. 104.

Description

The body whorl is large, but some what compressed on the side away from the aperture. The surface of the shell is interstices between the self grooves being flattened, and not raised into definite ridges. The outer lip of the aperture is greatly thickened and its inner edge bears a few crudely developed wart-like tubercles towards its inner margin adjoining the aperture, however, is deeply denticulate in the middle The shell is grayish black, profusely dubbed with white all over, the columella and the marginal surface of the outer lip are white and glossy

Scientific Name

Natica lamarckii

Classification

Phylum	:	Mollusca
Class	:	Gastropoda
Sub Class	:	Prosobranchia
Order	:	Archaeogastropoda
Super Family	:	Neritidae
Family	:	Neritidae

 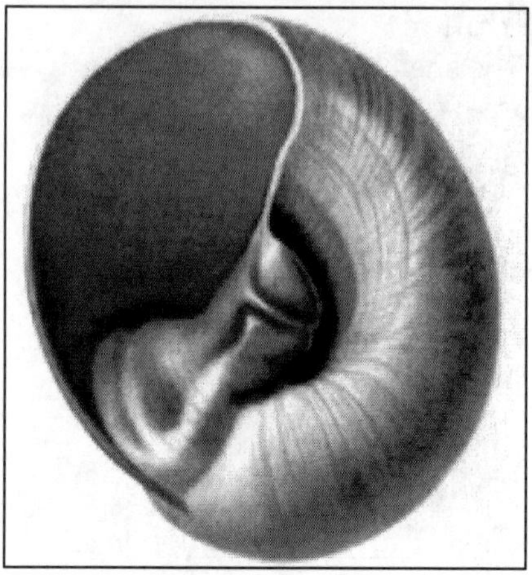

Sub Family : Neritinae

Genus : Nerita

Species : *lamarckii*

Description

Shell is large, thick, globosely with partially covered umbilicus, spire convex and little exerted; whorls rounded, concave in the upper part, body whorl inflated. Columella bears broad transversely expanded callus directed towards body whorl with a median groove but the anterior part of callus here as large as posterior one or little large and also the tip recurved. Umbilicus greatly covered leaving an angular opening. Surface is smooth, obliquely striated.

Size

Height 13.90 - 24.40 mm, maximum diameter 21.85 - 36.35 mm, minimum diameter 17.50 - 29.75 mm. Colour: Grayish white to pale brown.

Scientific Name

Natica lineate, Grevely

Classification

Phylum : Mollusca

Class : Gastropoda

Sub Class : Prosobranchia

Order : Mesogastropoda

Super Family : Naticacea

Family : Naticidae

Sub Family : Naticinae

Genus : Natica

Species : *lineata*

Reference

Natica lineate, Gravely, Bull. Mad. Govt. Mus. (Nat. Hist.) V. no. 2 1942.

Description

The shell is moderately large, with an inflated body whorl, a slightly elevated spire and a more or less sharply pointed, apex; it is comparatively this, with a smooth. and finely glossy surface when fresh. The aperture is large and the operculum is thin, but provided with a stout callosity which enters spirally inwards into the umbilical depression. The most characteristic feature of the shell, however, is its distinctive colour pattern, it is shy white, marked throughout with close set, transpired, wary, deep yellow or orange lines, the whorls of the spire are often partially tinged with pale bluish grey. Three shells are represented in the collection, two of which are rather faded, which all have the edge of the outer lip slightly broken.

Scientific Name

Nerita planospira Anton, 1839

Classification

Phylum	:	Mollusca
Class	:	Gastropoda
Sub Class	:	Prosobranchia
Order	:	Mesogastropoda
Super Family	:	Naticacea
Family	:	Naticidae
Sub Family	:	Naticinae
Genus	:	Natica
Species	:	*planospira*

Description

Shell with depressed spire, width greater than height. Sculptured are with about 30 spiral ribs, sometimes alternating large and small. Columella is deck smooth, with single elongate ridge posteriorly and 3-7 denticles centrally. Outer lip smooth. Exterior is dull, black, with white or reddish-brown axial streaks, white sometimes predominating (forming mainly white shells); brown periostracum sometimes present. Columella is deck white with black patch at edge centrally, aperture white. Operculum is calcareous, smooth, black or grey.

Size

Up to 30 in height

Scientific Name

Nerita plicata Linnaeus, 1758

Classification

Phylum	:	Mollusca
Class	:	Gastropoda

Sub Class	:	Prosobranchia
Order	:	Mesogastropoda
Super Family	:	Naticacea
Family	:	Naticidae
Sub Family	:	Naticinae
Genus	:	Natica
Species	:	*plicata*

Fig. 2

Description

Shell with low spire, width equal to or less than height. Sculptured are with about 16 strong spiral ribs. Columellar deck lirate, columella convex, with three strong denticles centrally. Outer lip is with large denticle at top, another at bottom, with 3-4 elongate denticles between. Exterior dull, white, fawn or pink, ribs sometimes spotted with black. Columellar deck white, aperture white. Operculum is calcareous, brown.

Size

Up to 30 mm in height.

Scientific Name

Natica tigrina Gravely

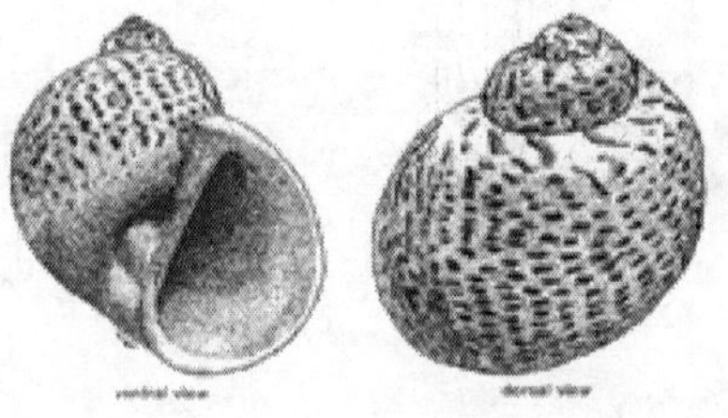
FAO

Classification

Phylum	:	Mollusca
Class	:	Gastropoda
Sub Class	:	Prosobranchia
Order	:	Mesogastropoda
Super Family	:	Naticacea
Family	:	Naticidae
Sub Family	:	Naticinae
Genus	:	Natica
Species	:	*tigrina*

Synonyms

Conchlis tigrina, Roading, Mus. Boltem. 1798, P/ 147

Natica maculosa, Lamarck, Anim. Sans. Vart., VIII, 1838, P. 641

Reference

Natica tigrina, Gravely, *Bull. Mad. Govt, Mus.(nat.Hist.)*, 1942, P 37.

Description

The shell is smaller in average size than that of *N. marochiensis* and may at once be distinguished from the rest by its characteristic colouration. The body whorl is inflated and globular, the general proportions of the shell and the shape of the umbilicus are very much the same as in *N. marochiensis*. The callus is very thick, especially at the base of the columella the shell is pale brownish, dotted all over with small, close-set, rounded purplish brown spots which are for the most part arrange in regular spiral rows coalescing with those of the adjacent one.

Scientific Name

Nerita rumphii Reeve

Classification

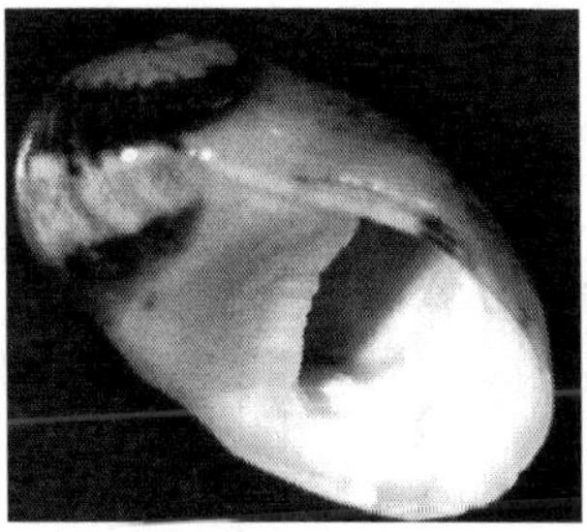

Phylum	:	Mollusca
Class	:	Gastropoda
Sub Class	:	Prosobranchia
Order	:	Archaeogastropoda
Super Family	:	Neritidae
Family	:	Neritidae
Sub Family	:	Neritinae
Genus	:	Nerita
Species	:	*rumphoo*

Synonyms

Nerita polita Var Rumphoo, Schepmen.

Reference

Nerita rumphii Reeve, *Conch. Icon*. IX, 1865, Nerita, Pl xi.

Description

The shell is moderate – sized, with a practically smooth and even glossy surface, through on close examination it is seen to be very finely Transpirally striated. The columellar surface is slightly concavely sunk and feebly ridged or granulated in the middle, and it's margin adjoining the aperture is be set with two or three weakly developed teeth. The outer lip is thickened and its marginal surfaces sloped steeply inwards, its inner edge faintly crenulated. This species is reported to be extremely variable in its colouration.

Scientific Name

Nassa arcularia

Classification

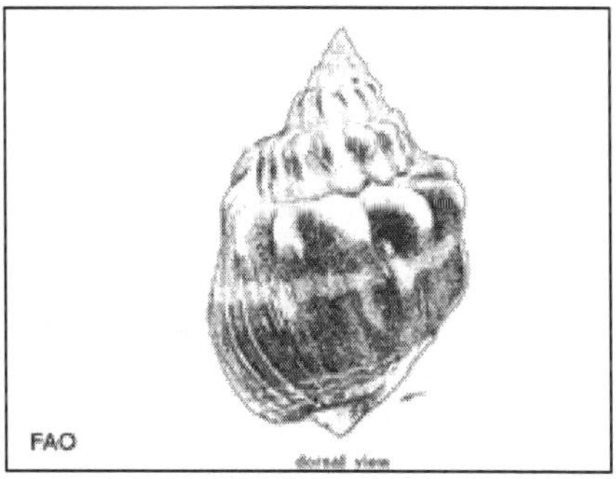

Phylum	:	Mollusca
Class	:	Gastropoda
Sub Class	:	Prosobranchia
Order	:	Neogastropoda
Family	:	Muricidae

Genus : Nassa

Species : *arcularia*

Description

The side of the columela is covered by a more or less broad and thick plate and the emarginated is deep but without a canal. The animal resembles a true buccium and there are gradual transitions among the shells from one subgenus to the other.

Scientific Name

Nassa hepatica Gravely

Classification

Phylum : Mollusca

Class : Gastropoda

Sub Class : Prosobranchia

Order : Neogastropoda

Family : Muricidae

Genus : Nassa

Species : *hepatica*

Synonyms

Buccinum monil, Kinerner, Con. Viv., IX, 1834 P. 48

Reference

Nass hepatica, Gravely, *Bull. Mad. Govt. Mus.(Nat.Hist.)* V No w, 1942, p.61

Description

The shell is larger, and bears a well elevated spire with a sharp apex. The entire surface is traversed by strong transpiral ribs which tend to be slightly flexuous on the body whorl the whorls bears a single strong spiral groove immediately below the suture cutting across the transpiral ribs at this part and thus giving rise to a series of nodules between it and the suture above, on the body whorl, however, there is a group of about four close-set spiral grooves at the base in addition to the single one above near the suture, the interior of the aperture is blackish brown.

Scientific Name

Nassarius (Zeuxis) dorsatus (Roeding) 1798

Classification

Phylum : Mollusca

Class : Gastropoda

Sub Class : Prosobranchia

Order : Meso gastropoda

Super Family : Nasariitea

Family : Nassaniidae

Sub Family	:	Nassarinae
Genus	:	Nassarius
Species	:	*dirsatus*

Description

Elongate-ovate, first 2-3 teleoconch whorls axially ribbed, remainder smooth and glossy, outer lip varixed behind but with a sharp or thin edge, finely dentate along its inner side, the denticles not extending into the aperture as lirate, columellar callus thick but narrow, continuous between the siphonal and anal canals, terminating with a prominent spine projecting over the siphonal canal.Size: Length 32.60-39.42 mm; diameter 18.22-21.95 mm; height of spire 17.50-21.30 mm; height of aperture 17.50-21.30 mm. Color: Variable, most commonly greenish grey or cream with brown dorsal blotches which may tend to form bands, sometimes uniformly brownish grey, tan, or even pink.

Scientific Name

Nassarius (Zeuxis) foveolatus (Dunker) 1847

Classification

Phylum	:	Mollusca
Class	:	Gastropoda
Sub Class	:	Prosobranchia
Order	:	Meso gastropoda
Super Family	:	Nasariitea
Family	:	Nassaniidae
Sub Family	:	Nassarinae
Genus	:	Nassarius
Species	:	*foveolatus*

Description

Shell larger in size, exceeding 12 mm in height; columella usually calloused, outer lip with denticles inside. Surface of the shell smooth, sculpture restricted to spiral whorls only; suture channeled.

Scientific Name

Nassarius fossatus Gould

Classification

Phylum	:	Mollusca
Class	:	Gastropoda

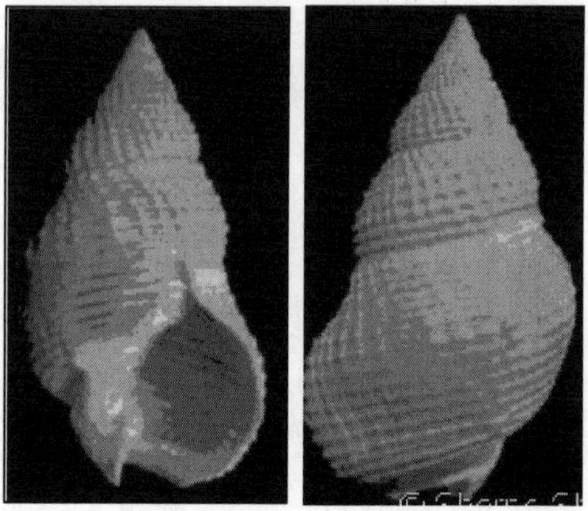

Sub Class : Prosobranchia
Order : Meso gastropoda
Super Family : Nasariitea
Family : Nassaniidae
Sub Family : Nassarinae
Genus : Nassarius
Species : *fossatus*

Reference

Nassarius fossatus, Gould A. A. 1849

Description

The shell is 1 1/2 to 2 inch in length, orange brown to brownish white in colour. Early whorls beaded, last whorl with about a dozen coarse, variously sized, flat topped spiral threads and with about a dozen short axial ribs on the top third of the last whorl. Outer lip is with a jagged edge and constricted at the top.

Scientific Name

Nassarius (Niotha) jacksonianus Quoy & Gaimard 1833

Classification

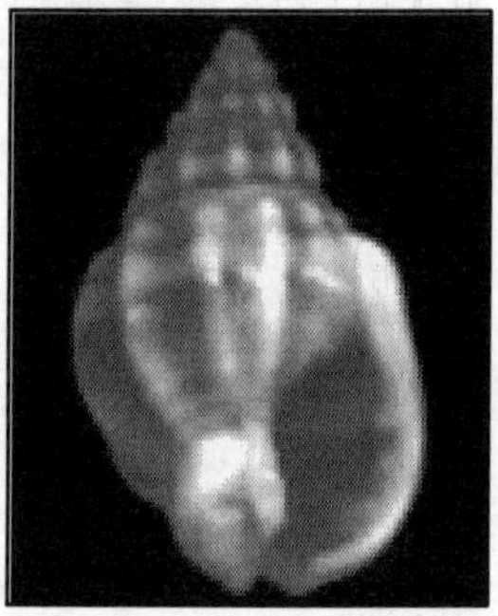

Phylum : Mollusca
Class : Gastropoda
Sub Class : Prosobranchia
Order : Meso gastropoda
Super Family : Nasariitea
Family : Nassaniidae
Sub Family : Nassarinae
Genus : Nassarius
Species : *jacksonianus*

Description

Spiral cords absent below suture; body whorl with faint bands on upper part only.

Size

Length 9.95-15.56 mm; diameter 4.95-9.05 mm

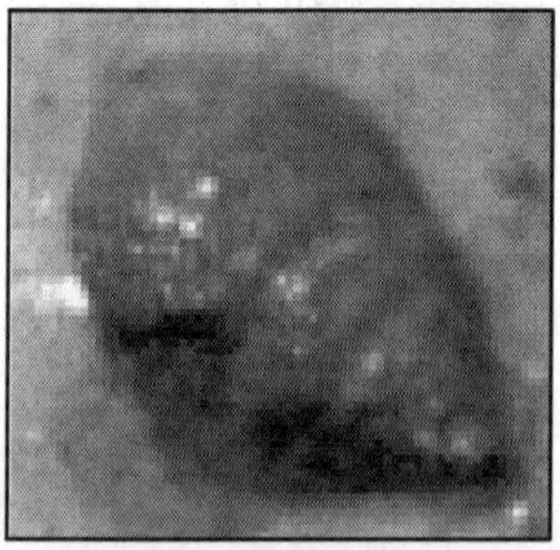

Scientific Name

Nassarius obsoletus

Classification

Phylum : Mollusca
Class : Gastropoda
Sub Class : Prosobranchia

Order	:	Meso gastropoda
Super Family	:	Nasariitea
Family	:	Nassaniidae
Sub Family	:	Nassarinae
Genus	:	Nassarius
Species	:	*obsoletus*

Description

The shell is 3/4 to 1 inch in length, usually curved with a mud and algae, and has its spire eroded the tip. Colour is dark black – brown. Sculpture of numerous rows found of weak beads. Parietal wall thickly glassed with brown and gray. Columella is with a single, strong spiral ridge near the base. Outer lip is with half a dozen small grayish teeth which run back into the aperture. Very common shell found on oozy, warm and flats.

Scientific Name

Nassarius trivittattus

Classification

Phylum	:	Mollusca
Class	:	Gastropoda
Sub Class	:	Prosobranchia
Order	:	Meso gastropoda
Super Family	:	Nasariitea
Family	:	Nassaniidae
Sub Family	:	Nassarinae
Genus	:	Nassarius
Species	:	*trivittattum*

Description

It is found in Gulf of Kutch (Positra, Narara). The shell is 3/4 inch in length. Wather light shelled, 8/9 whorls, nuclear whorls smooth. Whorls are in shire with 4 to 5 rows of strong, distinct beads. Partial wall thinly glazed with white enamel. Outer lip is sharp and thin. Whorls slightly channeled just below the suture. Colour light ash to yellowish grey. Commonly found form shallow water to 45 fathoms.

Scientific Name

Ocenebra circumtexta

Classification

Phylum	:	Mollusca
Class	:	Gastropoda
Sub Class	:	Prosobranchia

Order : Neogastropoda

Family : Sicioidae

Genus : Ocenebra

Species : *Circumtexta*

Description

The shell is 3/4 to 1 inch in length, spire 1/3 the length of the shell. Characterized by very strong, rough spiral cords 15 the cords are seen to consist of arched, crowded, raised brown. Axial ribs wide, low, rounded and 7 to 9 per whorl. Some specimens are banded or have large, red brown spots. A white form of this species was unnecessarily named *citrida Dall*. It is very abundant species on rocks at low tide to 30 fathoms.

Scientific Name

Oliva gibbosa (Born)

Classification

Phylum : Mollusca

Class : Gastropoda

Sub Class : Prosobranchia

Order : Neogastropoda

Family : Olividae

Genus : Oliva

Species : *gibbosa*

Description

This is well known and very conspicuous molluscs. In typical form be described as a broad massive shell a fine example measuring a good 3 ½ inch to over 3 ¾ inch longitudinally and by1 ½ inch in breadth. The normal colouration is being a livid mottled grey, much spotted or suffused with cancerous. When outer cuticle is removed the mottled surface shows much more clearly, the colour then being a fine chocolate or sepia. The white space are seen to be mostly narrow, long, often triangular or cuneiform or shaped like the letter L

The shell is spire and the whole of the collmeller region exhibit an unusual amount of callosity.

Scientific Name

Oliva reticularis

Classification

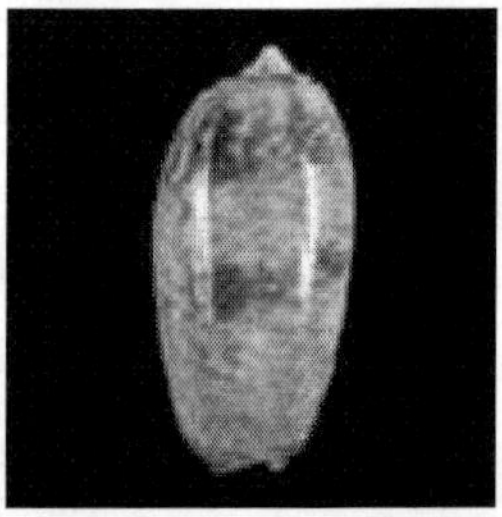

Phylum : Mollusca

Class : Gastropoda

Sub Class : Prosobranchia

Order : Neogastropoda

Super Family : Volutacea
Family : Olividae
Genus : Oliva
Species : *reticularis*

Description

The size of this animal is 1½ to 1¾ inches in length, similar to *Sayana*, but smaller more globose with an only finish and generally more lightly coloured. The golden olive or golden panama is merely a rate orange form of this species. Some times pure white or very dark brown in colour. It is commonly found in Gulf of Kutch.

Scientific Name

Oliva subulata Lian &C

Classification

Phylum : Mollusca
Class : Gastropoda
Sub Class : Prosobranchia
Order : Neogastropoda
Family : Olividae
Genus : Oliva
Species : *subulata*

Description

The shell is oblong or elliptical in shape. The aperture of which is narrowed long and emarginated appoisite to the spire, which is short. The folds of collumella are numerous and resemble striae. The whorls are sulciform. These shells are quite as beautiful as the cypraea.

Scientific Name

Potomides fascus

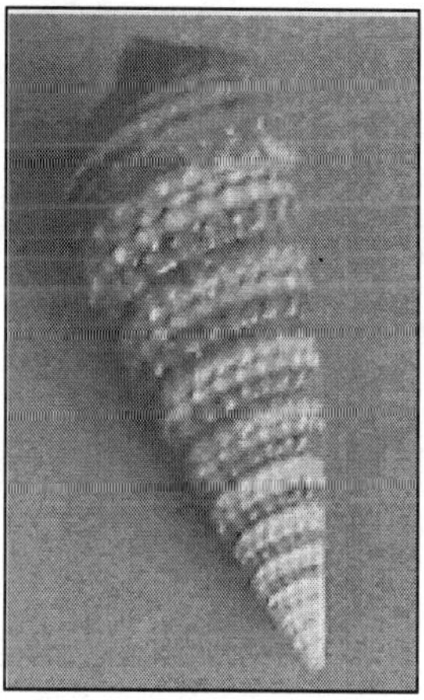

Classification

Phylum : Mollusca
Class : Gastropoda
Sub Class : Prosobranchia
Order : Meso gastropoda
Family : Potamidiae
Sub Family : Potamidinae
Genus : Potomides
Species : *fascum*

Description

The slender, brown-colored shells in the Potamididae family show an anterior notch near the top of the aperture, a feature shared in common with shells in the Cerithiidae family. However, the Potamididae shells have a less strongly developed anterior notch and a thin, horny operculum with many rather than fewer whorls. Like the Cerithiidae, the Potamididae, and Battillaridae molluscs are major dwellers in muddy fresh or brackish water. The Battilariidae differ from the Potanididae mainly in sperm structure. Shells generally range from 1 cm to 4 cm or more in length, considerably larger than most of the Cerithidae. The snails of the Cerithioidea superfamily generally have a tropical distribution.

At Narara, Poshitra, in Okha district, a large number of this species are found. Two large species the Horn Shells proper, belonging also to the genus potamides, and undoubtedly derived from the marine potomides are found in abundance crawling in the mud often out of water than in it. In the Philippine Island the flesh of the Horn Shells is stoned as food, the shells are thrown on wood fires and when sufficiently cooked, the apex of the spire is broken off and the animal sucked out through the broken end.

Scientific Name

Rapna bubosa Gravely

Classification

Phylum	:	Mollusca
Class	:	Gastropoda
Sub Class	:	Prosobranchia
Order	:	Neogastropoda
Super Family	:	Muricacea
Family	:	Thaididae
Genus	:	Rapana
Species	:	*bulbosa*

Synonyms

Buccinum bulbosum, Dillwayn Catalogue of shells, 1817, P. 631

Pyrula bulbosa, Reeve, Conch. Icon. IV, 1847, Pyrula. Pl iv

Reference

Rapana bulbosa, Gravely, *Bull. Mad. Govt. Mus. (Nat. Hist.)* V No 2. 1942, P.48

Description

The shell is large, with a very low spire and a broad, inflated body whorl. The surface is traversed. Through out by fine, close processes which extend even to the whorls of the spiral, in the body whorl, however, there may be one or two more incomplete taws of such tubercles below the principal one. The sutures are more or less markedly sunk in broad groves. The base of the body whorl is strongly narrowed. The umbilicus is very large and has a wide opening. The columellar surface is covered by a whitish, polished callus thickening. The aperture is very large and is produced into a short widely upon anterior canal. The surface of the shell is pale brownish throughout and the interior of the aperture white.

Scientific Name

Strombus urceus

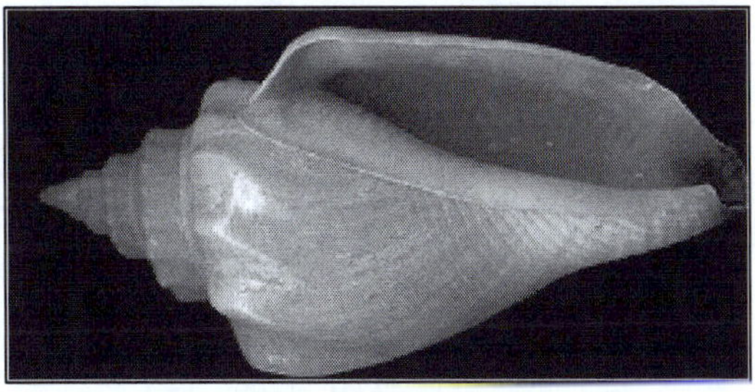

Classification

Phylum	:	Mollusca
Class	:	Gastropoda
Order	:	Neotaeioglossa
Family	:	Strombidae
Genus	:	Strombus
Species	:	*urceus*

Scientific Name

Solarisella obscura Couthoy

Classification

Phylum	:	Mollusca
Class	:	Gastropoda
Order	:	Archeogastropoda
Super Family	:	Trochoidae
Family	:	Trochidae
Genus	:	Solarisella
Species	:	*obscura*

Description

The shell is 1/4 inch in length, with whorls made more angular by one large, feebly beaded, spiral cord above the periphery. Base is smoothies except for microscopic spiral scratches. Umbilicus is narrower and bordered buy an angular rim. Colour grayish to pinkish tan, often worn to reveal a pearly golden colour. Aperture is pearly white. Some specimen may have weak aerial rib lets below the strongest spiral cord on the periphery of the whorl. They are commonly dredged out from 3 to 400 fathoms.

Scientific Name

Spiroglyphus lituellus Morch

Classification

Phylum	:	Mollusca
Class	:	Gastropoda
Order	:	Carenagastopoda
Family	:	Vermitidae

Genus : Spiroglyphus

Species : *lituellus*

Description

The shell is like small worm-tube molluscs found adhering to rocks and the shells of abalones in a tightly wound, flat spiral. The last whorl may grown up on top of the previous whorls and be erect for 1/4 of an inch. Aperture is circular, about 1/8 inch in diameter. Shell is solid, with two larger, scaled cords which give a somewhat squires cross section to the whole shell. Colour cream to purplish grey. Operculum is horny, multi spiral and brown moderately common.

Scientific Name

Succinea godivariana Rao

Classification

Phylum : Mollusca

Class : Gastropoda

Order : Systellammatophora

Family : Succineidae

Genus : Succinea

Species : *godivariana*

Description

The shell is small minutely rim ate, dispersedly sub vote, finely obliquely striated, pale greenish corneous spire very short, apex minutely papillae, suture deep. Aperture oblique, obviate, periscope simple, coloumella slightly thickened and refluxed, covering the exceeding minute, cubical chit, outer and basal margins regularly curved.

Scientific Name

Telescopium telescopium

Classification

Phylum : Mollusca

Class : Gastropoda

Order : Caenogastropos

Super Family : Cerithiacea

Family : Potamididae

Genus : Telescopum

Species : *telescopum*

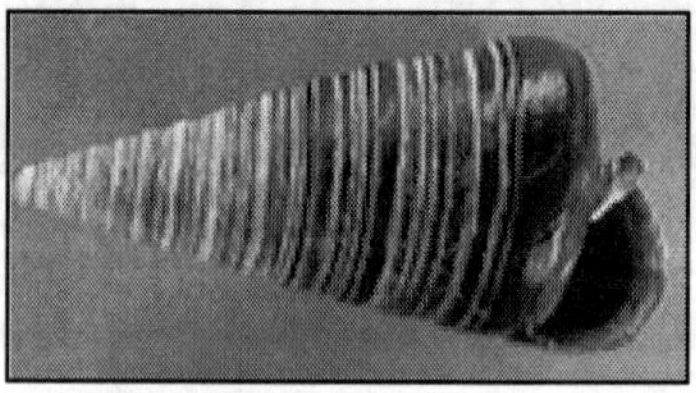

Description

A large, distinctive species with a thick, heavy shell, *T. telescopium* feeds on organic detritus and surface algae found on exposed mudflats. Their highly extendable snout can be seen in action in the photograph. Like the *mata merah*, this species also has a well-developed pallial eye. This animal is eaten locally.

Scientific Name

Territella cooperi Carpenter

Classification

Phylum	:	Mollusca
Class	:	Gastropoda
Sub Class	:	Prosobranchia
Order	:	Caenogastropoda
Super Family	:	Cerithioidea
Family	:	Turritellidae
Sub Family	:	Turritellinae
Genus	:	Territella
Species	:	*cooperi*

Description

The shell is 1 to 2 inches in length, 17 to 20 slightly convex whorls base concave. Columella and outer lip fairly fragile whorls with 2 or 3 small, spiral cords and usually with a number of much smaller, variously sized threads colour oranges to yellowish white with darker, axial flam mules. Moderately common just off shore.

Scientific Name

Thais bufo Gravely

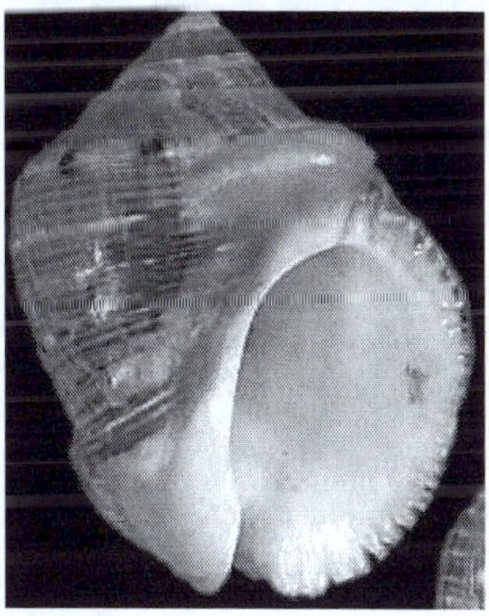

Classification

Phylum	:	Mollusca
Class	:	Gastropoda
Sub Class	:	Prosobranchia
Order	:	Neogastropoda
Super Family	:	Nuricacea
Family	:	Thaidiae

Sub Family : Thaidinae

Genus : Thais

Species : *bufo*

Synonyms

Purpura bufo, Lamarck, Ankm, Sans, Vert. X, 1845, P. 69

Reference

Thais bufo Gravely, *Bull. Mad. Govt. Mus. (Nat. Hist.),* V No 2, 1942.

Description

The shell is thick, large, and much broader than that of *T. rudolphi,* with a very low spire and more or less broadly inflated body whorl. The surface of the shell is uniformly spirally grooved throughout and on the body whorl there are two spiral rows of rather widely set, rounded tubercles, those of the upper row, whose site is marked by an obtusely angular shoulder, being the larger, below these rows of nodules there are two slightly squished spiral ridges, but they are not tunerculated. The aperture is broad with a slightly arched columellar border covered by a thick, polished callosity which is particularly prominent posteriorly. The outer lip is thick, finely crenulated within and deeply ovate. The shell is pale brown, faintly banded with dark brown ion the surface and the coumella and interior of the aperture are yellowish brown.

Scientific Name

Thais heptagonalis, Gravely

Classification

Phylum : Mollusca

Class : Gastropoda

Sub Class : Prosobranchia

Order : Neogastropoda

Super Family : Nuricacea

Family : Thaididae

Sub Family : Drupinae

Genus : Thais

Species : *heptergonalis*

Synonyms

Ricinula heptagonalis, Reeve, Conch. Icon., III, 1845, Riciniula, Pl.iii,

Reference

Thais margariticol Var. *Hepteagonalia,* Gravely, *Bull. Med. Govt. Mus. (Nat. Hist.),* No 2, 1942, P 50 and 53.

Description

The shell is readily distinguished by its brad, Transpiral ribs (of which there are not angularly shouldered as in the preceding species.). The spiral ridges are more uniformly developed, the main ridges being less sharply differentiated from the intervening ones. The inside of the outer lip bears a few, strong transversal elongated teeth. The shell is whitish irregularly marked with diffuse patches of dark or blackish brown. Generally disposed in the form of indistinct bands, the coloumella and interior of the aperture are pale violet.

Scientific Name

Thais sacellum (Gmelin, 1791)

Classification

Phylum	:	Mollusca
Class	:	Gastropoda
Sub Class	:	Prosobranchia
Order	:	Neogastropoda
Super Family	:	Nuricacea
Family	:	Thaidiae
Sub Family	:	Thaidinae
Genus	:	Thais
Species	:	*scellum*

Description

Shell massive, solid, of 6-7 whorls, of which the body whorl makes about two-thirds of total height. Whorls are with a distinct spiral keel bearing ragged, projecting knobs; and with squamose spiral cords, separated by intervals as broad as the cords. Body whorl is with 3-4 larger spiral cords on which the small cords are set close together, and the umbilical chink surrounded by a thickened structure. Aperture is ovate, terminating anteriorly in a small, short siphonal canal. Outer lip somewhat thickened inside, with 9-10 denticles extending far inside the aperture. Colour: outside of a yellowish tan colour, with dark brown mottles or flames over the spiral cords only, attenuated in the intervals. Inside of aperture is white. Common size: 30-40 mm.

Scientific Name

Tribia curate Sowerby

Classification

Phylum	:	Mollusca
Class	:	Gastropoda
Sub Class	:	Prosobranchia
Order	:	Orthrogastropoda
Super Family	:	Strombacea
Family	:	Strombidae
Genus	:	Tibia
Species	:	*curata*

Reference

Tibia curata, Sowerby, G.B.I, 1842

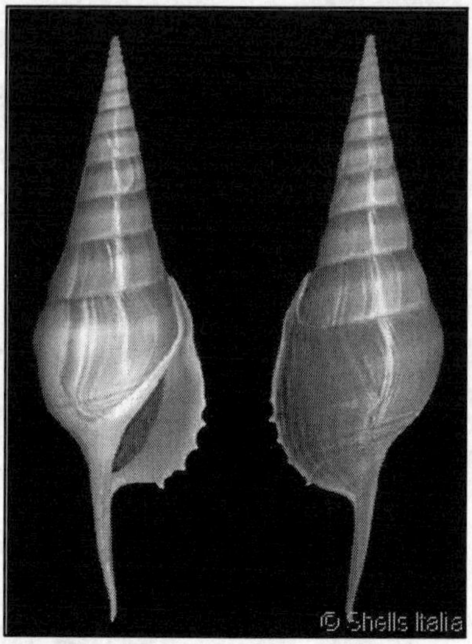

Description

The size of shell is 120 to 185 mm in length. The third Indian genus of the strombs is Tibia comprising the beak- shells. In these the shell has a finitely tapered spiral form, smooth and brown, the mouth whorl expands into a wide wing armed with several short and stout spines along the margin and with the anterior end prolonged into a slender, Semi-tubular rostrum or beak. Tibia curate is rare in the Gulf of Mannar but off Bombay and Malabar, Gujarat coast (Veraval, Okha, Porbander, Bhavnagar) and in the Gulf of Kutch area it is in large number. It is often abundant in depth between 10 and 30 fathoms. Animal is very like a strong but with the eyes near the base of the slender tentacles instead of being at the summit of long eye-stalk, the foot is often more conspicuously divided into this parts and equally obviously mount ended for scrambling over the ground at a rapid rat. The operculum is fan shaped and prettily sculptured with radiating lines.

Main Characteristics of Trochus

Have shells, in which the external margin of the angular aperture approaches more or less to a perfect quadrangular figure and is in an oblique plane with respect to the axis of the shell, because the part of the margin next to the spire projects more than the rest.

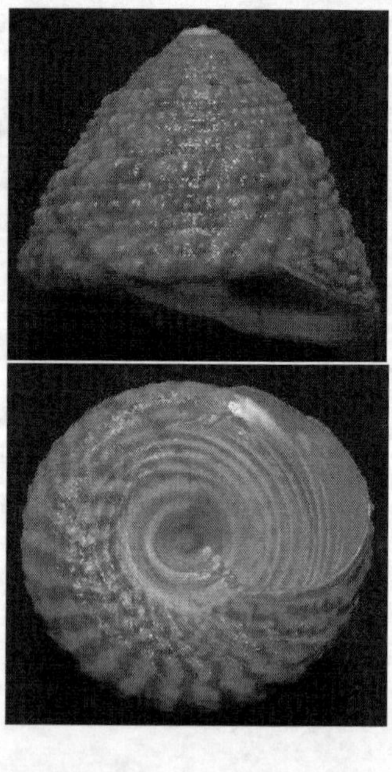

Scientific Name

Trochus radiatus Gamelin

Classification

Phylum	:	Mollusca
Class	:	Gastropoda
Sub Class	:	Prosobranchia
Order	:	Archaeogastropoda
Sub Order	:	Trochita
Super Family	:	Trochacea
Family	:	Trochidae
Sub Family	:	Trochinae
Genus	:	Trochas
Species	:	*radiatus*

Other Names

Trochas vividus Reve.

Reference

Trochus radiatus A. Adams, Proc. Zool. Soc. London. 1851. P. 185.

Description

The shell is moderate sized, slightly broader than high and of a more or less evenly conical shape. The surface of the whorls is traversed by regular, spiral rows of rounded tubercles in each whorl are larger than those of the lower rows. The basal surface of the body whorl is strongly spirally ridged. The columella is smooth and not denticulate. The shell is whitish or pale yellowish, brightly marked with broad, transpired, dark reddish bands which often get broken up into irregular spots on the basal surface of the body whorl.

Scientific Name

Trochus stellatus Gemlin

Common Name

Top Shell or Button shells

Classification

Phylum	:	Mollusca
Class	:	Gastropoda
Order	:	Vetigstropoda
Super Family	:	Trochoidae
Family	:	Trochidae
Genus	:	Trochus
Species	:	*stellatus*

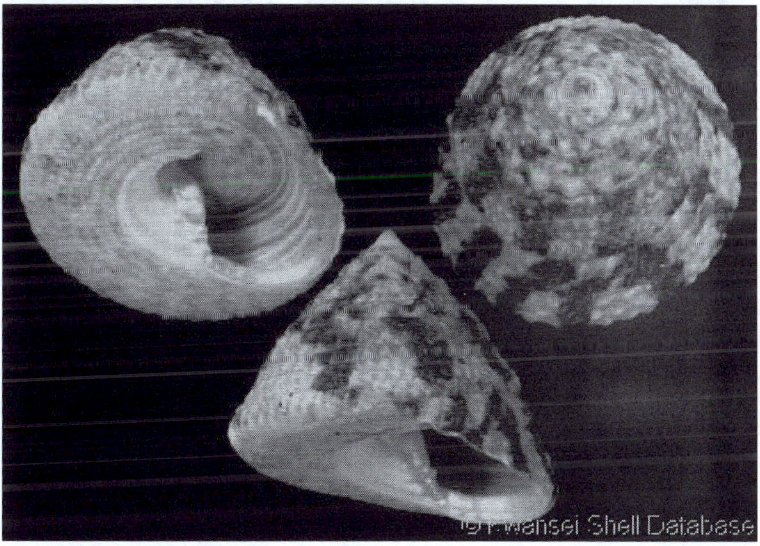

Description

The shell is 1 inch in length, with whorls made more angular by one large, feebly beaded, spiral cord above the periphery. Colour cream to purplish grey with chocolate bend on each whorls vertically. Operculum is horny, multi spiral and brown moderately common. Base is wide.

Scientific Name

Trochus tentorium Gamelin

Classification

Phylum : Mollusca

Class : Gastropoda

Sub Class : Prosobranchia

Order : Archaeogastropoda

Sub Order : Trochita

Super Family : Trochium

Family : Trochidae

Sub Family : Trochinae

Genus : Trochas

Species : *tentonium*

Synonyms

Trochus costatus, Gamlin

Polydenta tantorium, A. Adams

Reference

Trochus tentorium Gamelin, Syst., Nat., Ed. XIII, 1791, P. 3571, No 24.

Description

The shell may at once be distinguished from those of the preceding species by its more strongly nodular surface. The sides of the shell are more or less straight, much as in *T. rariatus*. The lower rows of tubercles in each whorl are enlarged as in *T. stellatus*, but they are much stronger and more pronounced and the stout fillet which they form is traversed by strong, thick, oblique, transpirally elongated ribs. The basal surface of the body whorl bears bended spiral ridges. The columella is distinctly denticulated. As in the preceding species, the shell is marked with transpire reddish bands, but they are much closer together in this species and often broken up into parches which are darker and more conspicuous in the lower part of each whorl.

Scientific Name

Turritella acropora Dall

Classification

Phylum : Mollusca

Class : Gastropoda

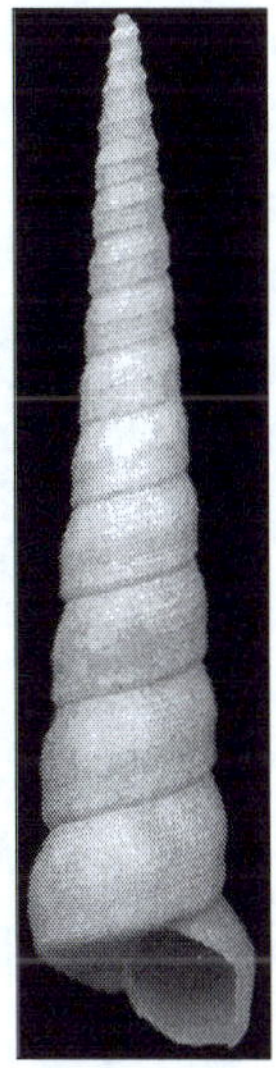

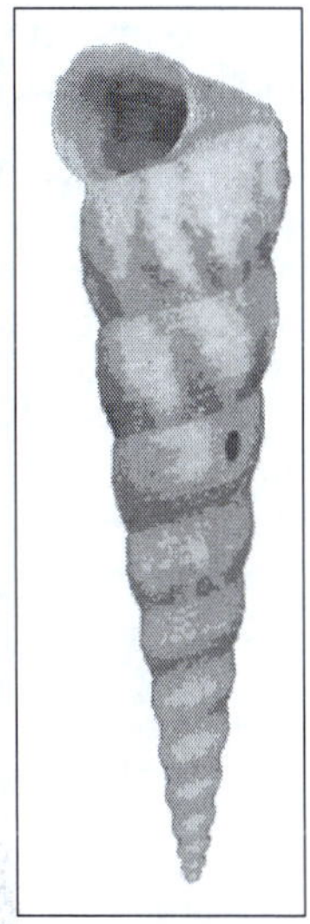

Sub Class : Prosobranchia
Order : Meso gastropoda
Super Family : Architeetonicacea
Family : Turritellidae
Sub Family : Turitellinae
Genus : Turritella
Species : *acropora*

Description

The shell is elongate, narrow, with 10-12 convex whorls. Aperture is squires. Color tan to brownish-orange. The shell is 1 inch in length resembling exocleta, but with convex whorls, and with numerous, fine, spiral threads a few of which, at the periphery, are slightly larger than the other. There is very weak series of driblets just below the suture. Colour yellowish to brownish orange. Commonly found in just off shore.

Main Characteristics of Turbo

The shell is round or oval the thick; the aperture is completed on the side next to the spire by the penultimate whorl. The animal has two long tentacles and the eyes placed on pedicle at their external base.

Scientific Name

Turbo argyrostomus

Classification

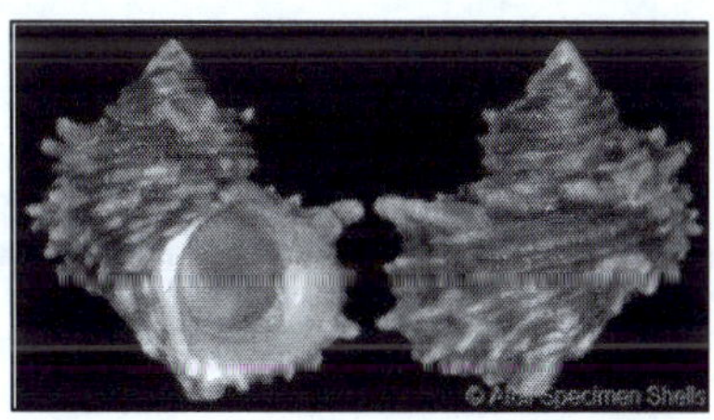

Phylum : Mollusca
Class : Gastropoda
Sub Class : Prosobranchia
Order : Archaeogastropoda
Sub Order : Trochina
Super Family : Trochicea
Family : Turbinidae
Sub Family : Turbininae
Genus : Turbo
Species : *argyrostomus*

Description

The larger kinds are commercially valuable, the widely spread 'Trocas' shell of trade and the even more valuable 'Green snail' of Eastern seas are in great demand for pearl button manufacture. The two families are easily distinguished by the nature of the operculum, a flattened hard structure found upon the upper surface of the tail end of the foot in many gastropods.

Scientific Name

Turbo castaneus Gaelin

Classification

Phylum : Mollusca

Class : Gastropoda

Sub Class : Prosobranchia

Order : Archaeogastropoda

Sub Order : Trochicea

Family : Turbinidae

Sub Family : Turbininae

Genus : Turbo

Species : *castaneus*

Description

Shell is 1 to 1½ inch in length. Colour of the shell found is oranges, greenish, brownish or grayish, commonly banded with flame-like white spots. Aperture is white callus on columella heavy. Lower lip projects downward. Operculum is calcareous.

Scientific Name

Tubo (Marmarostoma) intercostalis Menke 1843

Common Name

Turban shell

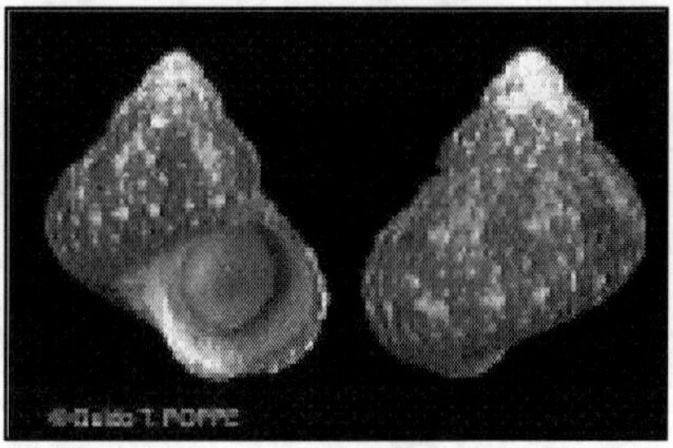

Classification

Phylum : Mollusca

Class : Gastropoda

Sub Class : Prosobranchia

Order : Archaeogastropoda

Sub Order : Trochicea

Family : Turbinidae

Sub Family : Turbininae

Genus : Turbo

Species : *intercostalis*

Description

Operculum is calcified and rather thick, with only a few spiral lines. Typically found in water with a depth of 0 to -5,467 meters (0 to -17,936 feet).

Scientific Name

Turbo margaritacus

Classification

Phylum : Mollusca

Class : Gastropoda

Sub Class : Prosobranchia

Order : Archaeogastropoda

Sub Order : Trochina

Super Family : Trochicea

Family : Turbinidae

Sub Family : Turbininae

Genus : Turbo

Species : *margaritaceus*

Description

This small species of turbo, allied to the well-known mother of pearl shell called "Green snail" in trade, is found in abundance of the surf washed rocky shoals found here and there on the coral reefs near Gulf of Kutch.

The island women collected the shells during spring tides when the shoals uncover for a short time, they employ all the available minutes in turning as many stones as possible, this being the favorite habitat of Turbo. The flesh is used in the same way as that of cockle-clam.

Beside the value of the flesh, Turbo is esteemed for its operculum. This is stout, dies-shaped stony body, flat on the inner surface, strongly convex on the outer, it serves as a defensive stopper when the animal is attached, the soft body being refracted whorly with in the snail-shaped.

Scientific Name

Turritella copperi Carpenter

Classification

Phylum : Mollusca

Class : Gastropoda

Sub Class : Prosobranchia

Order : Mesogastropoda

Super Family : Architectonicacea

Family : Turritellidae

Sub Family : Turritellinae

Genus : Turritella

Species : *copperi*

Description

The shell is 1 to 2 inch in length, 17 to 20 slightly convex whorl base concave. Columella and outer lip fairly fragile whorls with 2 or 3 small, spiral cords and usually with a number of much smaller, variously sized threads colour oranges to yellowish white with darker, axial flammables. Moderately common found just off shore.

Scientific Name

Umbonium vestiarium (Linne)

Classification

Phylum	:	Mollusca
Class	:	Gastropoda
Sub Class	:	Prosobranchia
Order	:	Orthrogastropoda
Sub Order	:	Vetigastropoda
Super Family	:	Trochoidae
Family	:	Trochidae

Genus : Umbonium

Species : *vestriarium*

Description

The shell is flattened, button-shaped, usually with a smooth and glossy surface. The umbilicus is filled up with a thick callus. The species is common on all Indian sandy shore. The shell is small and brightly coloured, with a smooth, highly polished surface, the spire is depressed, and the body whorl broad and more or less markedly flattened, and strongly angular below. The aperture is flattened and some what D-shaped. The umbilicus is filled up entirely by a thick, whitish callus which spreads over the basal surface of the shell. These molluscs live in the surface layers of sand in shallow water.

Scientific Name

Vexillum coroneunse J. cate

Classification

Phylum : Mollusca

Class : Gastropoda

Sub Class : Prosobranchia

Order : Neogastropoda

Genus : Vexillum

Species : *coronense*

Description

Shell is small, slender, bursiform, somewhat tubrculate, spiral about the same height as the body whorl. Protoconch lacking in both known specimens, tele-conch consists of 8 flatly convex whorls. Axial sculpture is of 10 prominent costae. Shell colour bright, deep orange except neck and colomella which are both porcelain-white. First 4 whorls bear narrow white spiral bank, final 2 to 3 whorls are surrounded by a fainter orange band at periphery. Neck of shell, starting at first posterior columellar fold, is coarsely ribbed, granulose. Labarum thin, simple constricted at base, with numerous faint irregular liras with in.

Scientific Name

Xancus pyrum Linne

English Name

Sacred Chank

Classification

Phylum : Mollusca

Class : Gastropoda

Sub Class : Prosobranchia

Order : Neogastropoda

Super Family : Volutacea

Family : Vasidae

Genus : Xancus

Species : *pyrum*

Description

The secret chank is exploited from the intertidal and reef habitat in the Gulf of Kutch. They are found on or about the coral reefs where one can wade through the shallow water to pick the chanks. Chanks prefer a depth of 10 – 27 meters with fairly high content of Calcium in the medium, a sandy bottom, with a mixture of mud in the association of tube dwelling polychaete worms, echinoderms, filamentous green algae, broken or loose corals. Normally when male attains 57 mm to 60 mm size, they are mature, while females are bigger than this when mature. Fertilization is internal and the fertilized ova are deposited in spindly twisted chitinous capsule. Such capsules contain about 28–36 eggs. The young ones develop in this enclosed capsule and when they are about 1 cm in length they are believed to be eating out the portion of the capsules and come out of the capsules.

Distribution

South East coast of India and north-west coast of India.

Area of Abundance

Gulf of Mannar, Palk Bay (Tamil Nadu coast) and Gulf of Kutch.

Chapter 4
Marine Scaphopoda

Introduction

SCAPHOPODA, the third of the five classes into which the Phylum Mollusca is divided. The Scaphopoda are marine Molluscs with the body, especially the foot, adapted to a burrowing life in sand. The structure is bilaterally symmetrical, the body and shell elongated along the antero-posterior axis and nearly cylindrical. The right and left margins of the mantle are united ventrally, leaving an anterior and posterior aperture to the mantle cavity. The shell has therefore the form of a tube open at both ends. The head is somewhat rudimentary and without eyes, but bears two dorsal appendages produced into numerous long filaments. Buccal mass and radular apparatus are present, but ctenidia are entirely wanting. The foot is cylindrical. At first supposed to be tubicolous Annelids, *Dentalium* and its allies were afterwards placed among the Gastropoda, to which recent authorities consider them to be closely related. In 1857 Lacaze-Duthiers raised them to the rank of a division equal to Lamellibranchia. This view is now generally adopted. The shell is narrower at the posterior end and is slightly curved to the dorsal side. Both the vernacular name, "tooth shell," and the Latin name, *Dentalium*, refer to the resemblance of the shell to a long tooth. The animal grows at the anterior end, and therefore the shell at the posterior end is older and thicker. The edge of the mantle at the anterior aperture is very thick and muscular; at the posterior aperture also there is a circular muscle, and here the edge is interrupted by a ventral sinus and is provided internally with a dorsal and ventral valve which can be applied to each other so as to close the aperture. The living animal buries itself in the sand with only the posterior extremity projecting into the Siphonopodiidae. At the base of the head dorsally are a pair of flat tentacular lobes from the edges of which the cephalic filaments or captacula arise. These captacula are of unequal length, highly contractile and extensile, easily thrown off and regenerated. They are ciliated, and their extremities are enlarged and have a small lateral depression in each. The captacula are tactile and prehensile and can be protruded from the anterior aperture of the mantle. The foot is elongated and cylindrical, and can be protruded from the anterior aperture to serve as a burrowing organ. In Dentaliidae it is pointed at the end and has an oblique projecting fold on either side behind the extremity. In Siphonopodiidae it ends in a disk with papillated margins, and in Pulsellum there is a filament in the centre of the disk. Two retractor muscles pass back from the base of the foot to the dorsal side of the shell.

Internal Anatomy

The cavity within the head leads into a true buccal cavity situated within the body at the base of the foot. This buccal sac is provided with a dorsal mandible and a ventral radula. The latter is short and carries five teeth in each transverse row. The intestine is short and forms several loops all situated close behind the foot. The stomach is small; into it open a small pyloric caecum and the ducts of the liver, paired in Dentaliidae, one on the left only in *Siphonodentalium.* The anus opens just behind the base of the foot. The liver is placed entirely behind the intestine in the middle of the body, and behind it the rest of the body is occupied by the unpaired gonad. The vascular system is very rudimentary. Heart and blood-vessels are entirely absent; the blood is contained in sinuses which have no distinct walls or endothelial lining, and the principal of which are the perianal, the pedal, the visceral and the pallial. It is remarkable that in Scaphopoda only among Mollusca the blood-spaces are in communication with the external medium: a pair of apertures near the renal openings led from the perianal sinus to the exterior and allow the blood to escape during violent contractions of the body. There are no special respiratory organs, their function is carried on by the internal surface of the mantle.

The nervous system resembles that of Gastropoda and Lamellibranchia. A pair of cerebral ganglia lies on the dorsal side of the oesophagus: they innervate the proboscis or head and its tentacular lobes and captacula. Close to each cerebral ganglion is a pleural ganglion, and each is connected by a long nerve with the pedal ganglion of the same side, the two connectives of either side being united in the distal part of their course. The pedal ganglia are situated in the middle of the foot. The pleural ganglia are also united by a long visceral commeasure as in Lamellibranchs, and this commeasure bears two ganglia lying close beneath the epidermis in front of the anus. There is also a stomatogastric system arising from the cerebral ganglia.

Eyes are absent; attached to the pedal ganglia are a pair of otocysts. They are innervated from the cerebral ganglia. The buccal cavity contains a sense-organ on the ventral side called the sub-radular organ. It consists of ciliated epithelium, beneath which are two ganglia connected with the labial commeasure by nerves. The only other sense-organs are the captacula, which are tactile and olfactory. Each contains a terminal ganglion connected with sensory cells in the lateral pit.

The sexes are separate. The gonad, whose position has already been mentioned, is divided into transverse lobes; its duct is anterior and single, and diverges to the right to open into the right kidney as in primitive Gastropods and Lamellibranchs.

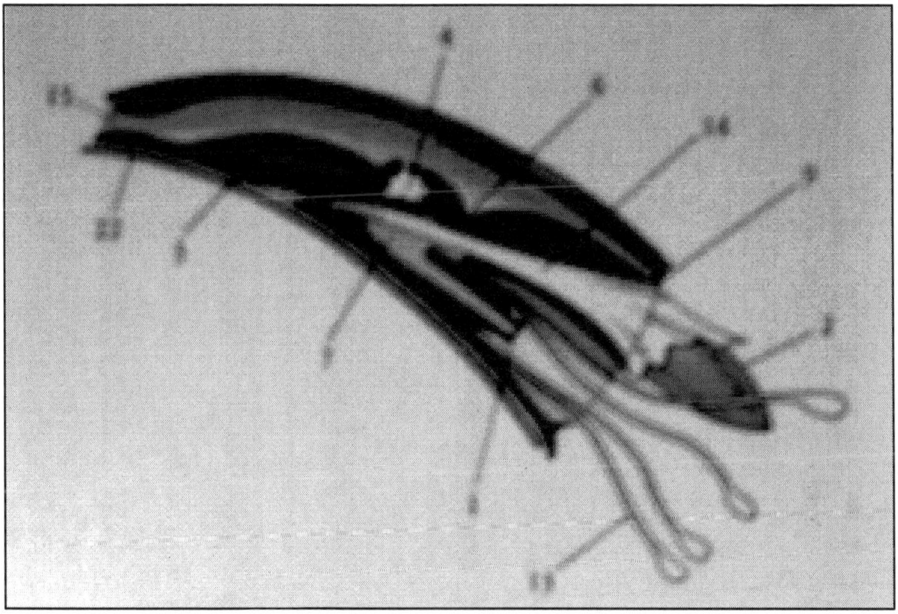

Anatomy of the Scaphopod

1: Mouth; 2: Foot; 3: Gonad; 4: Kidney; 5: Heart; 6: Anus; 7: Gut; 8: Gill; 9: Mantle; 10: Radula; 11: Operculum 12: Head tentacles; 13: Feeding tentacles; 14: Retractor muscle; 15: Inhalant; 16: Exhalant siphon; 17: Inhalant muscle; 18: Palps; 19: Adductor muscle; 20: Hinge; 21: Digestive gland; 22: Shell;

These are the tusk shell and are among the easier molluscs to place into their correct class. They are never found livings inter tidally but their shells are strong and so are frequently washed up on to the shore. They have white shells which are slightly curved, tapering, and tubes up to 120 mm long. They normally live head first (the wide end of the shell) in the sand and pump water into the mantle cavity through the shell aperture in the top of the shell. There are no gills and they feed on organic matter in the sand

Scientific Name

Dentalium aprinumn Linne

Classification

Phylum	:	Mollusca
Class	:	Scaphopoda
Order	:	Dentallida
Family	:	Dentalidae
Sub Family	:	Dentaliinae
Genus	:	Dentalium
Species	:	*octangulatum*

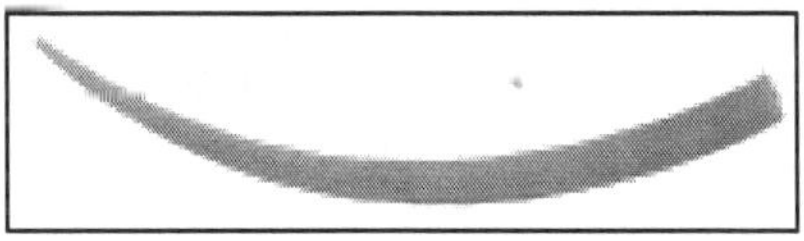

Reference

Dentalium aprinum, Linne, *Syst. Nat., Ed.* XII,1766, P. 1263

Description

The shell is rather long, slender and well arched, especially attaches posterior part. The length is about twelve times as much as the maximum diameter. The sculpture consists of about then narrow and sharply raised longitudinal ribs, separated by much wider, rather deeply concave interstices which are frequently divided by a five medium rib lets. The ribs are stronger on the concave side of the shell. Growth striate is inconspicuous as in the preceding species. The aperture is polygonal in cut line, the sides of the polygon, however, being concavely depressed. The ana aperture is minute circular, and not indented. The shell areas bleached white, but when fresh, the shell banded with pale green.

Scientific Name

Dentalium elips Winckworth, 1927

Classification

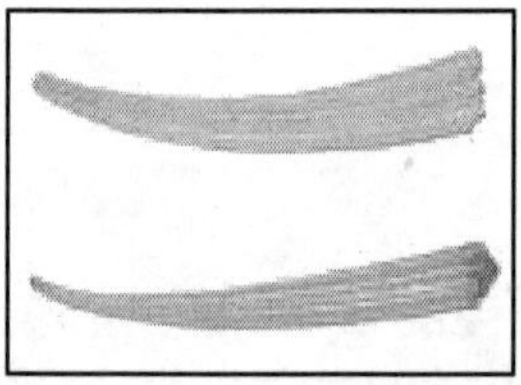

Phylum	:	Mollusca
Class	:	Scaphopoda
Order	:	Dentallida
Family	:	Dentalidae
Sub Family	:	Dentaliinae
Genus	:	Dentalium
Species	:	*elips*

Reference

Dentalium elips, Winckworth R. 1927. Marine Mollusca from India and Ceylon.-I. *Dentalium. Proceedings of the Malacological Society of London* 17 (4): 167-169, pl. 14.

Description

The shell is narrow markedly slender not much widened interiorly and only very slightly arched. The shell is readily distinguished from those of the preceding species by its peculiar sculpture. The surface is very finely and closely ribbed over a little more than half the length from the posterior end, while interiorly the shell is quite smooth. There are also fine, evenly developed growth lines which are visible under the lens on the smooth, glossy, anterior part of the shell. The shell is pale bluish white for the most part, but the posterior tip is more or less pure white.

Scientific Name

Dentalium octangulatum Donovan, 1803

Classification

Phylum	:	Mollusca
Class	:	Scaphopoda
Order	:	Dentallida

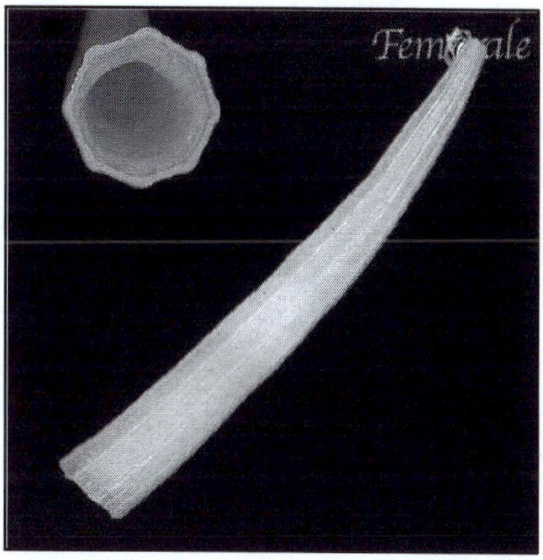

Family	:	Dentalidae
Sub Family	:	Dentaliinae
Genus	:	Dentalium
Species	:	*octangulatum*

Reference

Dentalium octangulatum, Hornel, common Molluscs of the South India, *Med, Fish. Bull.,* XIV.

Description

The shell is rather large, moderately stout and about 10 times as long as the diameter of the front ends. It is well curved. The sculpture consists normally of eight very strong, raised, rounded longitudinal ribs which are particularly sharp and prominent towards, the anterior aperture they often become slightly less strongly rose towards the anterior aperture, they are separated buy wide, fairly deep, more or less flattened or only slightly concave interstices towards the wider end, growth lines are in conspicuous. The shell is white or slightly bluish white and displays a fine gloss. In young specimens the shell is more strongly curved than in the adults, and the interstices between the ribs are smooth.

FRESHWATER MOLLUSCS SHELLS

Chapter 5
Freshwater Molluscs of India

FRESHWATER GASTROPODA

Introduction

Freshwater molluscs–snails and bivalves–have been used frequently as bioindicator organisms. With increasing needs for research on contaminant effects in freshwater ecosystems, this kind of biomonitoring is likely to develop further in the future. Molluscs can be used effectively for studies of both organic and inorganic contaminants; this review focuses on studies involving bioaccumulation and toxicity of metals. Two important advantages of snails and bivalves over most other freshwater organisms for biomonitoring research are their large size and limited mobility. In addition, they are abundant in many types of freshwater environments and are relatively easy to collect and identify. At metal concentrations that are within ranges common to natural waters, they are generally effective bioaccumulators of metals. Biomonitoring studies with freshwater molluscs have covered a wide diversity of species, metals, and environments. The principal generalization that can be drawn from this research is that bioaccumulation and toxicity are extremely situation dependent; hence, it is difficult to extrapolate results from any particular study to other situations where the biological species or environmental conditions are different. Even within one species, individual characteristics such as size, life stage, sex, and genotype can have significant effects on responses to contaminants. The bioavailability of the metal is highly variable and depends on pH, presence of organic ligands, water hardness, and numerous other controlling factors. Despite this variability, past studies provide some general principles that can facilitate planning of research with freshwater snails and bivalves as metal bioindicators. These principles may also be useful in understanding and managing freshwater ecosystems.(Reviews of Environmental Contamination and Toxicology; (United States); Journal Volume: 122)

Diagnostic Shell Features

Different reproductive biologies and certain anatomical adaptations for reproduction from major bases for classification of freshwater clams above the level of species, whereas species, and especially those in the same genus, are recognized by shell characters.

Classification

Freshwater molluscs, being thoroughly polyphyletic, cannot be classified independently of the rest of the great invertebrate phylum mollusca. Species are grouped (classified) into different higher categories, *e.g.*, genera and families, on the bases of natural affinities. All species of freshwater clams were described from shell characteristics, and the vast majority of groups were also originally established according to shells. It must be realized, however, that classification is not a static exercise. Reclassification occurs as new discoveries about organisms are made.

Importance to Man

The importance of freshwater molluscs to humans and the environment is greater than their general absence in collections and popular guides would lead one to believe, although with increasing environmental awareness, and threats of extinction, this lately has started to change. Some negative aspects, where they are a source of problems themselves, will be considered first:

Parasite Vectors

While landsnails and slugs may play minor roles in the transmission of certain tapeworms and nematodes, the importance of freshwater molluscs as intermediate hosts for a great variety of human and animal parasitic flatworms cannot be ignored. Prosobranch gastropod taxa known to be involved in parasite life cycles include *Pila*, snails belonging to the Hydrobiidae and related families, Thiaridae including *Thiara*, Pleuroceridae including *Elimia* (*Juga*), and *Semisulcospira*, and also several marine gastropods in the related family Potamididae. Pulmonate gastropod hosts include the Planorbidae, Bulinidae (now included within the Planorbidae), and the Lymnaeidae (Chandler and Clark, 1961).

Scientific Name

Ariophanta laevipes Muller

Classification

Phylum	:	Mollusca
Class	:	Gastropoda
Sub Class	:	Pulmonata
Order	:	Basommatophora
Family	:	Apiophantidae
Sub Family	:	Macrochlamydinae
Genus	:	Ariophanta
Species	:	*laevipes*

Reference

Ariophanta laevipes, Tonapi and Leela Mulherkar, *J. Bom.Nat. Hist.* 1963, Vol. 60(1), p. 116.

Description

The shell is relatively depressed, rather thin, with coloured bands. The spire is low, and there are five whorls which are more or less convex. The body whole has an angulated periphery but is rounded below. The apex is depressedly conoidal. The aperture is very oblique and almost diagonal to the axis. The peristome is moderately thickened and reflected below. The lower part of the body whorl is smooth. The ground colour of the shell is variable from white to brown or dark brown. But the species is characterized by three spiral chestnut bands, one close to the suture, and one above and one below the periphery. The parietal walls of the aperture and the area surrounding the umbilicus have the same colour as the body whorl.

Scientific Name

Bithynia stenothyroides Dohrn

Classification

Phylum	:	Mollusca
Class	:	Gastropoda
Sub Class	:	Prosobranchiata
Order	:	Megagastropoda
Family	:	Hydrobiidae
Genus	:	Bithynia
Species	:	*stenothyroides*

Reference

Bithynia stenothyroides, Tonapi and Leela Mulherkar, *J. Bom.Nat. Hist. 1963,* Vol. 60(1), p.108.

Description

The shell is small ovate globosely with four to five whorls. The last whorl is strongly inflated and is larger than all the other together. The whorls are convex with the spire conical. The aperture is broadly oval with the lip continuous but feebly reflected. The shell is semi transparent, glossy with a faint bluish tinge. The shell is devoid of any conspicuous sculpture. The operculum is concentrically striated.

Scientific Name

Cyclophorus (Annularia) aurantiacus Schumacher

Classification

Phylum	:	Mollusca
Class	:	Gastropoda
Sub Class	:	Prosobranchiata

Order : Megagastropoda

Family : Cyclophoridae

Genus : Cyclophorus

Species : *aurantiacus*

Reference

Cyclophorus aurantiacus Tonapi and Leela Mulherkar, *J. Bom.Nat. Hist.* 1963, Vol. 60(1) p. 106.

Description

The shell is quite large, thick and solid. The spire is turbinate, more depressed and relatively broad. The last whorl is broad and the whorls of the spire are convex. All the whorls are transpirally striated but the basal part of the body whorl is smooth and fine. The conspicuously large aperture is circular and has a thickened peristome which is reflected out. The shell is fulvous white with zig- zag deep brown marks. The interior of the apertural lip is bright orange while the inner portion is whitish.

Scientific Name

Cryptozona (Xestina) belangeri Var bombayana Pfeiffer

Classification

Phylum : Mollusca

Class : Gastropoda

Sub Class : Pulmonata

Order : Basommatophora

Family : Apiophantidae

Sub Family : Macrochlamydinae

Genus : Cryptozona (Xestina)

Species : *belangeri*

Reference

Cryptozona (Xestina) belangeri, Tonapi and Leela Mulherkar, *J. Bom. Nat. Hist.* 1963, Vol. 60(1), p.115.

Description

The shell is large, more or less globose, with depressed and low spire. The whorls are slightly inflated and distinctly convex with well impressed sutures. The aperture is roundly lunette, not as broad as high. The surface of the shell is obliquely striated with decussated impressed lines which are sometimes absent. The basal region is relatively smooth. The colour is extremely variable, ranging from dull white, pale horny, and sometimes to tawny brown.

Scientific Name

Cyclophorus (Litodylus) involvulus Muller

Classification

Phylum : Mollusca
Class : Gastropoda
Sub Class : Prosobranchiata
Order : Megagastropoda
Family : Cyclophoridae
Genus : Cyclophorus
Species : *involvulus*

Reference

Cyclophorus Involvulus, Tonapi and Leela Mulherkar, *J. Bom.Nat. Hist.* 1963, Vol. 60(1), p.105.

Description

The shell is turbinate and the spire is elevated with inflated whorls. The shell is spirally striated and striae on the body whorl from spiral ridges. The aperture is oblique to the axis and is circular. The peristome is thick and markedly reflected. The umbilicus is broad and deep but partially occluded by the reflected perisotme. The shell is chestnut coloured with brownish white wavy marks.

Scientific Name

Cyclophorus indicus Deshayes

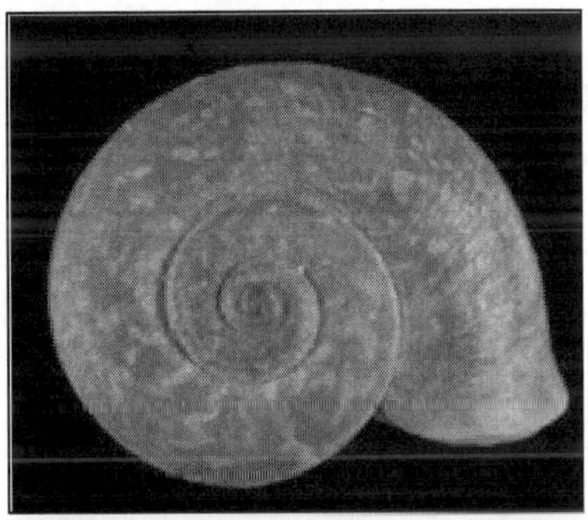

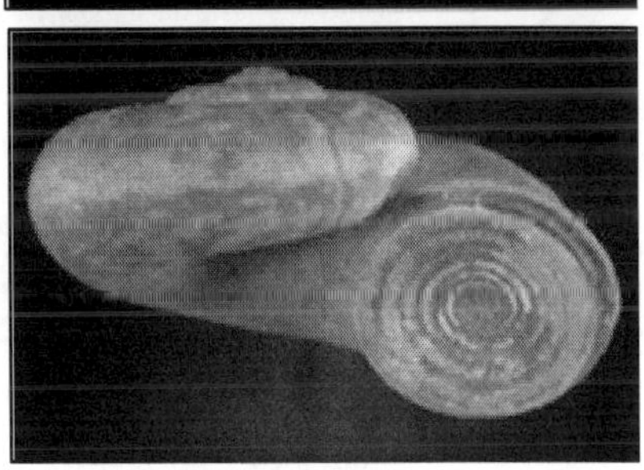

Classification

Phylum : Mollusca
Class : Gastropoda
Sub Class : Prosobranchiata
Order : Megagastropoda
Family : Cyclophoridae
Genus : Cyclophorus
Species : *indicus*

Reference

Cyclophorus Indicus., Tonapi, G. T. *J. Bom. Nat. Hist. Soc.* 1971. Vol. 68(1), 116

Description

Shell orbicular, turbinated, with an acute apex, Spire conical and composed of 6 – 7 convex whorls, Sutures hardly impressed, Body whorl distinctly inflated bearing a projecting keel in the circumference. Umbilicus is very narrow and deep. External surface of shell covered, particularly near the sutures, thick and thin transpiral striations. Circular aperture oblique to the axis with both lips thickened and reflected. Shell, fawn coloured but the lower part often mottled with dark brown flames. Banding is conspicuous on the body whorl and surrounding the umbilicus.

Scientific Name

Digoniostoma pulchella Benson

Classification

Phylum : Mollusca

Class : Gastropoda

Sub Class : Prosobranchiata

Order : Megagastropoda

Family : Amnicolidae (Hydrobiidae)

Genus : Digoniostoma

Species : *pulchella*

Reference

Digoniostoma Pulchella, Tonapi, G. T., 1971, J. Bom. Nat. Hist. Soc., Vol. 68(1), 117.

Description

The shells are broadly and irregularly ovate with apex slightly compressed. There are about 4-5 whorls of which the first two are not so conspicuous. Body whorl rather inflated. Sutures oblique linear and well impressed. Aperture small, oblique ovately rounded. Outer lip thickened and produced; its extremities sub-angulated. Coloumellar callus thickened and laminated in appearance. A distinct projection is developed where the columellar callus meets the lip. Umbilicus almost closed. The channel running forward from it is not well formed and is smaller.

Scientific Name

Fanus ater Linne

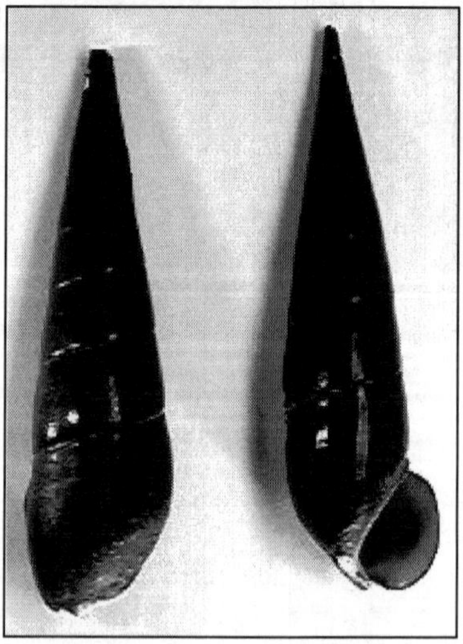

Classification

Phylum : Mollusca

Class : Gastropoda

Sub Class : Prosobranchiata

Order : Megagastropoda

Family : Melaniidae

Sub Family : Melanopsinae

Genus : Fanus

Species : *ater*

Reference

Fanus ater Tonapi and Leela Mulherkar, *J. Bom.Nat. Hist.1963, Vol. 60(1)* p. 109

Description

The shells are elongatedly tapering and turreted with the whorls more or less flatted. The apex is acuminate and the sutures are well impressed. The lower whorls have widely spaced spiral grooves mixed with dense transpiral striations on the surface. The aperture is small, ovately angled above but with a broad basal anterior canal. The labrum is sharp and thin. The surfaceis rough with iron rust colour and some specimens are bleached into yellowish olive-brown colour. The numbers of whorls are eleven to twelve with the 12 mm diameter major, apertureal height 10 mm and apertureal width 5.6 mm.

Scientific Name

Glessula notigena Benson.

Classification

Phylum : Mollusca

Class : Gastropoda

Sub Class : Prosobranchiata

Order : Megagastropoda

Family : Glessulidae

Genus : Glessula

Species : *notigena*

Description

The imperforate shell is elongated, conical with an attenuated apex. Spire slender, turetted with obtuse apex. Sutures rather deeply impressed with about 9=10 whorls which are feebly convex. The body whorl is not so tumid. The last whorl is 1/3 of the rest. Aperture rather oblique an semiovate. The peritremal margin is acute and thin while the columellar margin and the columella are curved and truncated obliquely near the base. The shells are horny brown glossy with crenulations particularly near the sutures. Some have even a fulvous horny colour.

Scientific Name

Lymnaea acuminate Lamarck

Classification

Phylum : Mollusca

Class : Gastropoda

Sub Class : Pulmonata

Order : Basommatophora

Family : Lymnaeidae

Sub Family : Lymnaeinae

Genus : Lymnaea

Species : *acuminata*

Reference

Lymnaea acuminate Tonapi and Leela Mulherkar, *J. Bom.Nat. Hist.1963, Vol. 60(1)* p. 111

Description

The shell is ovately oblong, smooth, thin and semi translucent. The body whorl is slightly angular above and inflated below the middle. The spire is short, narrow, with the attenuated whorls forming a pointed apex. The aperture is wide with the columellar lip twisted. There is a fine close-set transpiral striation on the surface of the shell. The body whorl is clearly demarcated from the spire with an abruptly narrowed base.

Scientific Name

Lymnaea luteola Lamarck

Classification

Phylum	: Mollusca
Class	: Gastropoda
Sub Class	: Pulmonata
Order	: Basommatophora
Family	: Lymnaeidae
Sub Family	: Lymnaeinae
Genus	: Lymnaea
Species	: *luteola*

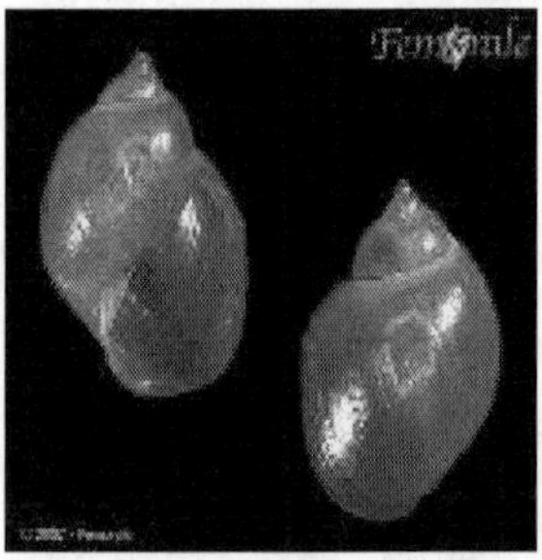

Reference

Lymnaea luteola, Tonapi and Leela Mulherkar, *J. Bom.Nat. Hist.* 1963, Vol. 60(1) p. 111.

Description

The shell is broader in proportion to the height in this species than in *L. acuminate*. The spire is short, conical, with a broader base which gradually merges with an inflated body whorl. The whorls of the spire are clearly inflated. To ovate aperture is rather angularly narrowed above but is rounded below. The shell is smooth, glossy, with a pale yellow horny tinge. The sculpturation consists of close-set fine transpiral striations, which are seen only under binocular microscope. In addition to these a few widely spaced spiral striae were also seen. The colour of the columellar fold is opaque white and is a rather characteristic feature of this species.

Scientific Name

Macrochlamys pedina Benson.

Classification

Phylum	: Mollusca
Class	: Gastropoda
Sub Class	: Pulmonata

Order	:	Basommatophora
Family	:	Apiophantidae
Sub Family	:	Macrochlamydinae
Genus	:	Macrochlamys
Species	:	*pedina*

Reference

Macrochlamys pedina, Tonapi and Leela Mulherkar, *J. Bom.Nat. Hist.* 1963, Vol. 60(1) p.114..

Description

The genus *Macrochlamys* is represented by a little over hundred species in this continent and it is difficult to recognize the different forms as the difference in the shells are so small indeed, that they had to be arranged according to locality. The shell is dressed, very thin, and translucent. The spire is canoed but very low. The whorls are six and a half to seven in number and are slightly convex above. The body whorl is bluntly sub-angulate above the periphery but rounded below. The lunately round aperture is oblique to the axis. The peristome is thin, while the columellar lip is vertical ad reflected to cover part of the umbilicus. The umbilicus is wide and conspicuous. The shell is more or less smooth with minute close set transpiral striations. The colour is variable from pale yellow to fulvous horny.

Scientific Name

Melania (Plotia) scabra Muller

Classification

Phylum	:	Mollusca
Class	:	Gastropoda
Sub Class	:	Prosobranchiata
Order	:	Megagastropoda
Family	:	Melaniidae
Sub Family	:	Melaniinae
Genus	:	Melania (Plotia)
Species	:	*scabra*

Reference

Melania (Plotia) scabra Tonapi and Leela Mulherkar, *J. Bom.Nat. Hist.* 1963, Vol. 60(1) p. 109.

Description

Quite apart from the other characteristic which separate this species from the succeeding one it is broader in proportion to height than *M. tuberculata*. The whorls of the spire bear spinous shoulders in their upper part. The whorls also bear well developed transpiral ridges with spiny process. The aperture is ovate. The shell is variable in its colouration but usually is brownish olive, spotted with few rust-coloured transpiral marks. The apical region is often darker than the body and the penultimate whorl. The prominent angular ridges and the spinous processes are often worn out in older speciments.

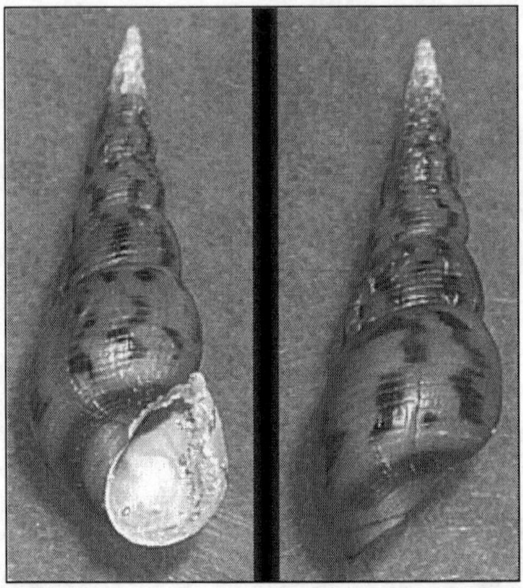

Scientific Name

Melania (Striatella) tuberculata Muller.

Classification

Phylum	:	Mollusca
Class	:	Gastropoda
Sub Class	:	Prosobranchiata
Order	:	Megagastropoda
Family	:	Melaniidae
Sub Family	:	Melaniinae
Genus	:	Melania (Striatella)
Species	:	*tuberculata*

Reference

Melania (Striatella) tuberculata Tonapi and Leela Mulherkar, *J. Bom.Nat. Hist.* 1963, Vol. 60(1) p. 109.

Description

The shell is elongated with an acuminate apex. The whorls are neatly convex and progressively increase towards the body whorl. The aperture is oval but more narrowed above and broadly rounded below. The sculpture consists of transpiral tuberculated ridges with raised spiral ridges. The body whorl is usually devoid of transverse tuberculated ridges. The colour of the shell is dark brown with rows of reddish undulating flame shaped discontinuous bands. The interior of the shell is glossy with external marks faintly visible.

Scientific Name

Paludomus (Stomatodon) stomatodon Benson.

Classification

Phylum	:	Mollusca
Class	:	Gastropoda
Sub Class	:	Prosobranchiata
Order	:	Megagastropoda
Family	:	Melaniidae
Sub Family	:	Paludominae
Genus	:	Paludomus
Species	:	*stomatodon*

Reference

Paludomus stomatodon Tonapi and Leela Mulherkar, *J. Bom.Nat. Hist.1963, Vol. 60(1)* p. 109

Description

The shell is very thick and solid and has a neritoid form and appearance. The shell has a depressed spire with a very strongly inflated body whorl. The aperture is sigmoidly oval with a tooth like projection in its basal margin. The apices of the spire are usually worn out or damaged. The shell is dark olive brown but the apex is black.

Scientific Name

Pisidium (Cyclocalyx) adamsi Prime

Classification

Phylum	:	Mollusca
Class	:	Gastropoda
Sub Class	:	Pulmonata
Order	:	Basommatophora
Family	:	Sphaeriidae (= Pisidiidae)
Sub Family	:	Pisidiinae
Genus	:	Pisidium
Species	:	*adamsi*

Description

Shell longer in relation to height, and of variable shape (sub trapezoidal to ovate); external surface usually glossy, with yellow, green or copper background color and numerous radiating green lines. Shell lacking a ridge atop each beak. External surface of the shell with fine concentric growth striate (about 15 per mm in the middle of the shell).

Scientific Name

Planispira proxima Ferussac

Classification

Phylum	:	Mollusca
Class	:	Gastropoda
Sub Class	:	Pulmonata
Order	:	Basommatophora
Family	:	Pleurodontidaii
Genus	:	Planispira
Species	:	*proxima*

Reference

Planispira proxima, Tonapi and Leela Mulherkar, *J. Bom.Nat. Hist.1963, Vol. 60(1)* p.117.

Description

The shell is moderate in size (Height – 12 mm, and diameter- 21 mm), conoidly depressed, and is rather globose. The whorls are five to five and a half in number, including rapidly in size until the body whorl is dilated near the aperture. The whorls are convex and sutures are deeply impressed. The apex is obtuse as the spire is depressed. The body whorl is slightly inclined towards the aperture, which is somewhat oblique to the axis and is broadly ovate. The margins of the aperture approach towards each other and are united by a thin callus on the parietal wall. The peristome is thickened, expanded, and reflucted. The sculpturation consists of close set, rather coarse, oblique transpiral striations. The colour of the shell is light fawn with a brown band at the periphery.

Scientific Name

Planorbis (Indoplanorbis) exustus Deshayes

Classification

Phylum	:	Mollusca
Class	:	Gastropoda
Sub Class	:	Pulmonata
Order	:	Basommatophora
Family	:	Planorbidae
Genus	:	Planorbis (Indoplanorbis)
Species	:	*exustus*

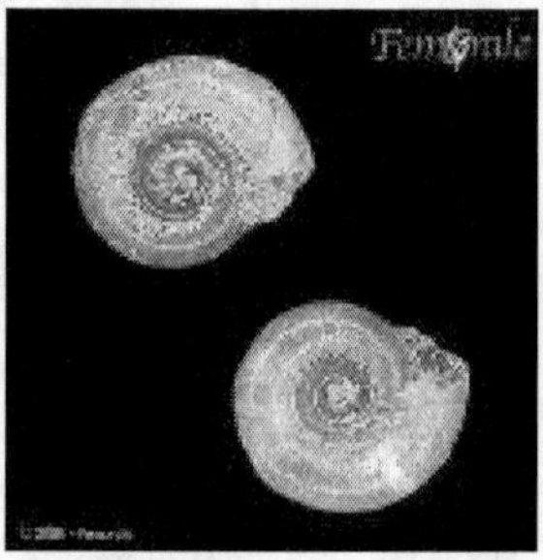

Reference

Planorbis (Indoplanorbis) exustus Tonapi and Leela Mulherkar, *J. Bom.Nat. Hist.* 1963, Vol. 60(1) p. 112.

Description

The species is extremely variable in its size (major diameter 15mm), form of its spire, and other architectural details. The shell is relatively large, moderately thick, and flattened on both sides. The shell is usually sinistral, discoidal, with the spire depressed and sunk in the expanded body whorl. The three whorls are quite convex and are spirally coiled in the horizontal plane. The outer lip is relatively thickened and feebly reflected outwards. The shell is transpirally and finely stirated throughout and the body whorl shows more distinct transpiral ridges. The range of variation in colour is considerable.

Scientific Name

Pila globosa Swainson

Classification

Phylum	: Mollusca
Class	: Gastropoda
Sub Class	: Prosobranchiata
Order	: Megagastropoda
Family	: Ampullaridae
Genus	: Pila
Species	: *globosa*

Reference

Pila globosa Tonapi and Leela Mulherkar, *J. Bom.Nat. Hist.* 1963, Vol. 60(1), p. 107.

Description

This species, with a globose shell, is most familiar. The spire is depressed but the whorls are inflated. The surface is very smooth and glossy. The aperture is oblong oval with the margin slightly thickened. The umbilicus is small, contracted by the raised peristome which is slightly reflected. The colour of the shell is brownish olive with irregular red-brown conspicuous spiral bands. The interior of the shell is shiny with yellow tinge and transverse reddish bands. The transpiral growth striae occur on the last whorl and are stronger near the aperture.

Scientific Name

Sulcospira hugely Var *compacta* Nevill

Classification

Phylum	: Mollusca
Class	: Gastropoda
Sub Class	: Prosobranchiata
Order	: Megagastropoda
Family	: Melaniidae
Genus	: Sulcospira
Species	: *hugely* Var *compacta*

Reference

Sulcospira hugely Var compacta, Tonapi and Leela Mulherkar, *J. Bom.Nat. Hist.* 1963, Vol. 60(1), p. 108.

Description

This is shorter and stouter variety of *S. hugely*. The body whorl is markedly angular and the spire is more truncate. The upper whorls are often found missing in older shells. The aperture is more or less contracted with the basal margin markedly produced below. The outer lip is sharp and thin. The columellar margin is smooth and white. The characteristic smoky-brown colour is darker in the dead shells. The spiral sulcations at the base of the body whorl are conspicuous.

Scientific Name

Vivipara bengalensis Lamarck

Classification

Phylum	: Mollusca
Class	: Gastropoda
Sub Class	: Prosobranchiata
Order	: Megagastropoda
Family	: Viviparidae
Sub Family	: Viviparinae
Genus	: Vivipara
Species	: *bengalensis*

Reference

Vivipara bengalensis Tonapi and Leela Mulherkar, *J. Bom.Nat. Hist.* 1963, Vol. 60(1).

Description

The shell is ovately conical. The lower part is more or less ovoid with a spire which is broadly conical. The whorls are rather inflated and the sutures are well impressed. The aperture is mango-shaped, *i.e.* angularly pointed above and rounded below. The sculpture consists of close-set fine transpiral lines and minute punctures. The coloration varies a great deal but sually the shell is olive-green with alternating broad and narrow dark brown spiral band. The narrower bands are lighter in colour than the broad ones. The former alone are present in the basal part of the body whorl.

Scientific Name

Vivipara dissimilis Muller

Classification

Phylum	: Mollusca
Class	: Gastropoda
Sub Class	: Prosobranchiata
Order	: Megagastropoda
Family	: Viviparidae
Sub Family	: Viviparinae
Genus	: Vivipara
Species	: *dissimilis*

Reference

Vivipara dissimilis Tonapi and Leela Mulherkar, *J. Bom.Nat. Hist.* 1963, Vol. 60(1), p.106.

Description

This is also common snail. It is treated by some authors as a variety of *V.bengalensis*. The shell is broader with a body whorl which is more ovoid. The sculpture consists of close-set delicate spiral striae and oblique transpiral growth striae which are prominent and well marked in the peripheral region of the body whorl. The ovoid aperture is covered by a horny operculum which is narrowed above. The umbilicus is narrower in this species than in *V. bengalensis*. The shell is dirty olive-green; the interior is of dull bluish white colour. This species is more in flowing water.

Scientific Name

Zootecus chin Pfeiffer

Classification

Phylum	:	Mollusca
Class	:	Gastropoda
Sub Class	:	Pulmonata
Order	:	Basommatophora
Family	:	Rumininae
Genus	:	Zootecus
Species	:	*chin*

Reference

Zootecus chin, Tonapi and Leela Mulherkar, *J. Bom.Nat. Hist.* 1963, Vol. 60(1) p. 114.

Description

The shell is of moderate size (Height – 12mm), smooth, and glossy. The spire is elongated and terminates in a conical apex. There are seven whorls, of which the body whorl is large and is approximately one-third the total length and is well rounded below. The penultimate and ante penultimate whorls are almost as broad as the body whorl. Thus the shell has a characteristic cylindrical pupiform appearance. The aperture is semi-oval with inner lip slightly reflected to partly occlude the umbilicius. The peristome is slightly thickened, with callus. The surface of the shell is covered by close set transpiral striae.

FRESHWATER PELECYPODA (BIVALVIA)

Introduction

Freshwater bivalves are bivalve molluscs which live in freshwater, as opposed to saltwater. Although the majority of species of bivalve molluscs live in the sea, a number of different families of bivalves live in freshwater (and in some cases in freshwater and brackish water). These families are grouped into two separate lineages which are not closely related.

Habitats for freshwater bivalves vary from very small ditches and ponds, to lakes, canals, and rivers.

One of the largest species of freshwater bivalves is the swan mussel, in the family Unionidae; it can grow to a length of 20 cm, and usually lives in lakes or slow rivers. Freshwater pearl mussels are well-known as a source of freshwater pearls.

The word "mussel" is also used for many freshwater bivalves, including the freshwater pearl mussels. Freshwater mussel species inhabit lakes, ponds, rivers, creeks, canals, grouped in a different subclass, despite some very superficial similarities in appearance.

The mussel's external shell is composed of two hinged halves or "valves". The valves are joined together on the outside by a ligament, and are closed when necessary by strong internal muscles. Mussel shells carry out a variety of functions, including support for soft tissues, protection from predators and protection against desiccation.

The shell has three layers. In the pearly mussels there is an inner iridescent layer of nacre (mother-of-pearl) composed of calcium carbonate, which is continuously secreted by the mantle; the prismatic layer, a middle layer of chalky white crystals of calcium carbonate in a protein matrix; and the periostracum, an outer pigmented layer resembling a skin. The periostracum is composed of a protein called conchin, and its function is to protect the prismatic layer from abrasion and dissolution by acids (especially important in freshwater forms where the decay of leaf materials produces acids).

Like most bivalves, mussels have a large organ called a foot. In freshwater mussels, the foot is large, muscular, and generally hatchet-shaped. It is used to pull the animal through the substrate (typically sand, gravel, or silt) in which it lies partially buried. It does this by repeatedly advancing the foot through the substrate, expanding the end so it serves as an anchor, and then pulling the rest of the animal with its shell forward. It also serves as a fleshy anchor when the animal is stationary.

Freshwater mussels inhabit permanent lakes, rivers, canals and streams throughout the world except in the Polar Regions. They require a constant source of cool, clean water. They prefer water with a substantial mineral content, using calcium carbonate to build their shells.

Scientific Name

Corbicula regularis Prime

Classification

Phylum	: Mollusca
Class	: Pelecypoda (Bivalvia)
Order	: Eulamellibranchiata
Sub Order	: Heterodonta
Family	: Corbiculidae
Genus	: Corbicula
Species	: *regularis*

Reference

Corbicula regularis, Tonapi and Leela Mulherkar, *J. Bom.Nat.Hist.* 1963, Vol. 60(1), p.119.

Description

The shell is triangular, ovate, transverse, and equilateral. The anterior and posterior margins are similarly rounded. Externally the shell is strongly and concentrically striated. The periostracum is dark brownish green while the interior is violet and glossy. The hinge bears three divergent cardinal teeth which are well developed in each valve. The lateral teeth are elongated, lamelliform, and bear fine transverse straie. The pallial line is continuous and is distinct with a shallow pallial sinus.

Scientific Name

Lamellidens corruanus Lea

Classification

Phylum	:	Mollusca
Class	:	Pelecypoda (Bivalvia)
Order	:	Eulamellibranchiata
Sub Order	:	Scizondonta
Family	:	Unionidae
Genus	:	Lamellidens
Species	:	*corrianus*

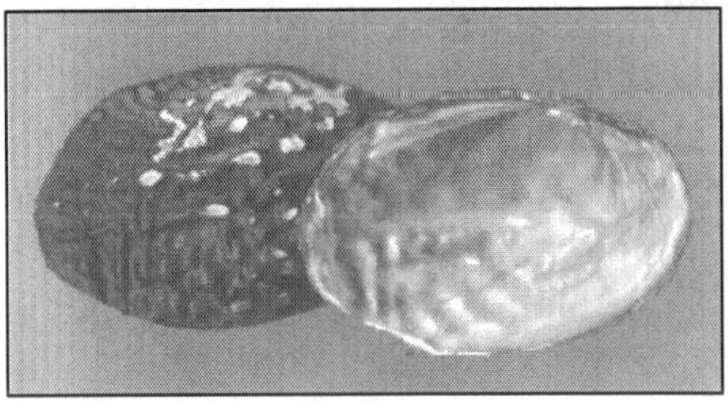

Reference

Lamellidens corrianus, Tonapi and Leela Mulherkar, *J. Bom.Nat.Hist.* 1963, Vol. 60(1), p.118.

Description

The shell is narrowly elliptical, strongly transverse, and is longer in proportion to the height (Length – 74 mm, height – 37 mm). The valves are relatively thin. The anterior margin is rounded, while the posterior margin is sub-angular. The beaks are not so prominent and usually the periostracal layer is eroded in fully grown shells. The ventral margin is roundly curved. The cardinal teeth, single in the left valve and paired in the right valve, are thin and bladed. The lateral teeth are elongated and nearly straight. The pearly white iridescence of the nacreous layer is characteristic.

Scientific Name

Lamellidens marginalis Lamarck

Classification

Phylum	:	Mollusca
Class	:	Pelecypoda (Bivalvia)
Order	:	Eulamellibranchiata

Sub Order : Scizondonta

Family : Unionidae

Genus : Lamellidens

Species : *marginulis*

Reference

Lamellidens marginulis, Tonapi and Leela Mulherkar, *J. Bom.Nat.Hist.* 1963, Vol. 60(1), p. 118.

Description

The shell is oblongly ovate, approximately twice as long as high, (Length- 76 mm, and height – 40 mm) and relatively more inflated. The umbones are prominent. The anterior margin is narrower than the posterior, the former evenly rounded while the latter is roundly angular. The shell is thin and covered with blackish brown, or greenish brown epidermis which is frequently worn away in the adult shells and more particularly near the umbone. The inner surface is iridescent and pearly.

Scientific Name

Parreysia corrugate Var *nagpoorensis* Lea

Classification

Phylum : Mollusca

Class : Pelecypoda (Bivalvia)

Order : Eulamellibranchiata

Sub Order : Scizondonta

Family : Unionidae

Genus : Parreysia

Species : *corrugate* Var *nagpoorensis*

Reference

Parreysia corrugate Var nagpoorensis, Tonapi and Leela Mulherkar, *J. Bom.Nat. Hist.* 1963, Vol. 60(1), p. 117.

Description

The shell is of large size (Length 45 mm and Height 22 mm), transversely, ovate, sub triangular and in-equilateral. The shell is moderately inflated and the valves are almost as long as high. The anterior end is narrowly rounded, while the posterior end is more or less broadly angular above and below. The prominent umbo is slightly inclined forwards. The periostracum is refocus -brown tinged with green towards the ventral margin. The umbonal region is ornamented with divaricating obliquely radial ridges. The hinge margin is slightly inclined downwards in front of and behind the umbo. The inner surface of the shell is pearly, slightly pinkish, and brilliantly iridescent. The cardinals are strong, jagged and crenulated. The laterals are lamellar, elongated and slightly curved.

Chapter 6
Land Molluscs

Land Snails

Snails and slugs (part of the group known as molluscs) are nature's recyclers, digesting material from the forest floor or the microscopic material that collects on foliage to release nutrients for use by other living things. The land snails are a large assemblage and there is a wide variety of diets amongst them. Many feed on decomposing material, algae, leaves, mosses or fungi but there is a carnivorous group which prefers to dine on other snails and invertebrates such as earthworms.

The mouth of a snail contains a small pad covered with tiny, serrated 'teeth' (called a radula) which enables the snail to scrape off matter as if it was using sandpaper to do the job. The head and foot are continuous and move forward with the aid of glands which produce a layer of mucous over which the foot muscle slides. A few families have an operculum which is stuck to the top of the foot when the foot is extended. When the snail retreats into its shell, the operculum becomes a door which closes fast, sealing moisture inside.

There is a group called semi-slugs which don't have an operculum or even much of a shell - their shells being more like 'hats' on top of their bodies. And lastly, there are the slugs which have no shells at all. If they encounter dry conditions, all they can do is shrink the body into a more compact shape and tuck their antennae under to reduce the amount of skin surface that is exposed to the air. The rainforest environment is normally a very favourable place to support a variety of slugs and semi-slugs.

Land Snail

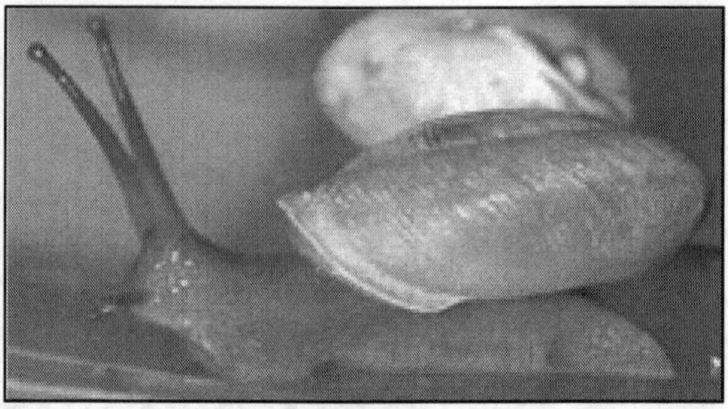

Land Snail

Semi Slug

Identification Key for Land Snails

1a. The top surface of the slug is uniform in pattern, *not* appearing to be divided into front and back parts *(Philomycidae and Testacellidae; the mantle covers most or all of the slug's top surface)*.

- Go to 2

1b. The top surface of the slug is apparently divided into two parts – front and back *(Arionidae, Limacidae, Milacidae; the slugs whose mantle covers only part of the top)*.

- Go to 6

2a. The top surface is heavily mottled or spotted, adults are often large *(more than 5cm)*.

- Go to 3

2b. The top is blue-gray, tan or very lightly speckled, and adults are often small *(5cm or less; Pallifera species)*.

- Go to 5

2c. The top is pale, with a tiny shell on the posterior *(Family Testacellidae).*

- *Testacella haliotidea* – earshell slug *(non-native)*

3a. The fringe or "sidewall" of the foot is pink or orange and the mucus is orange *(these are much easier to see in live animals)*. The top surface has one or two close gray-black central stripes.

- *Philomycus togatus* – toga mantleslug

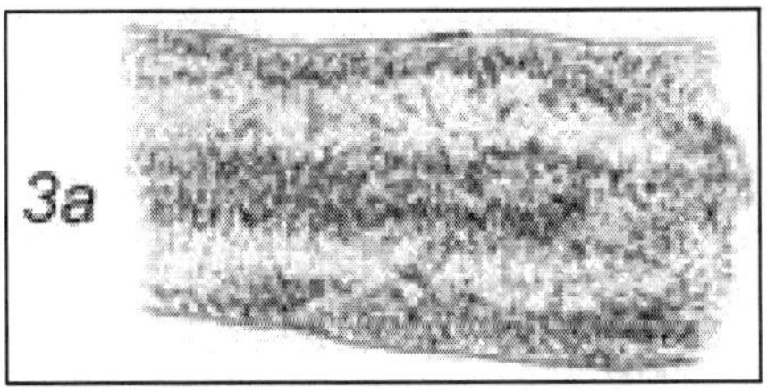

3b. The foot fringe is olive or gray *(this is a dark color that runs the entire length of the animal – don't be fooled by a dark tail tip, which is not diagnostic).*

- *Megapallifera mutabilis* – changeable mantleslug

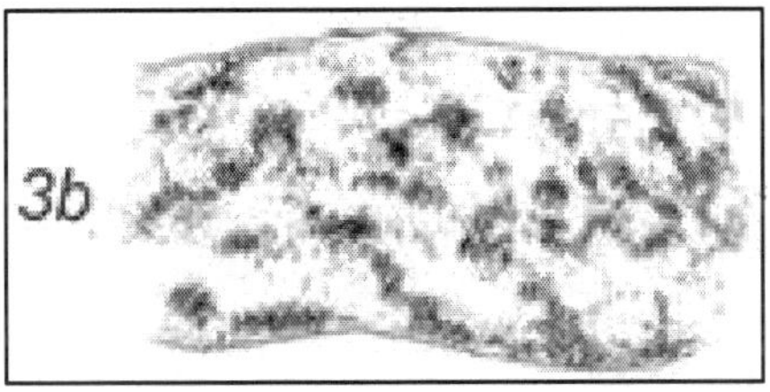

3c. The foot fringe is pale.

- Go to 4

4a. There is a central mottled stripe on top, plus a mottled stripe on each side *(the top stripe may sometimes appear as two very close stripes).*

- *Philomycus flexuolaris* – winding mantleslug

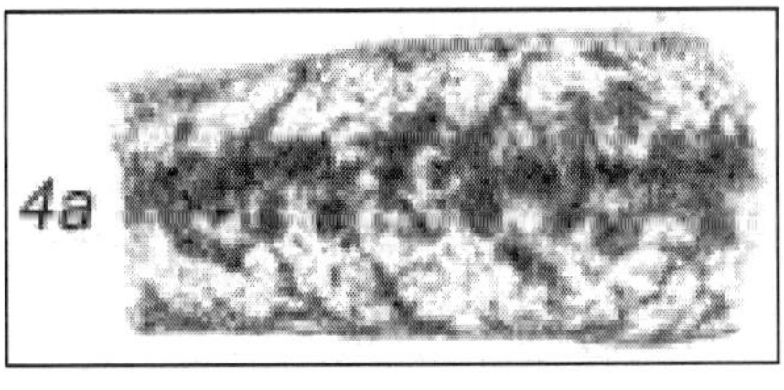

4b. There are two sets of dark, elongated spots along top, superimposed upon a paler, broad central stripe.

- *Philomycus carolinianus* – Carolina mantleslug

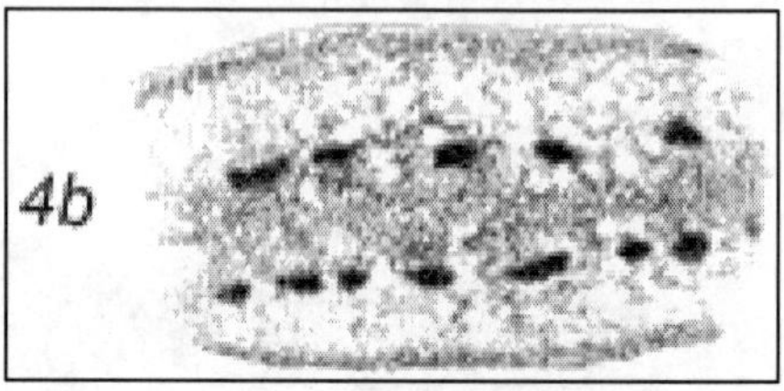

5a. The slug has a uniformly pale foot. There is a gray back, with a speckled center line most distinct near head.

- *Pallifera dorsalis* – pale mantleslug

5b. The sole of the foot has two narrow blood-red lines running lengthwise along edge, starting from front. The animal has a plain tan or gray back.

- *Pallifera ohioensis* – redfoot mantleslug

6a. The slug's top surface has a "keel" – an acute top edge – on at least the posterior part of the back. Posterior end of tail is sharp in top view and lacks a mucus pit. A "fingerprint" channel pattern on the mantle may be present or absent *(Family Milacidae, Family Limacidae).*

- Go to 7

6b. The slug has a keel only on the tip of the tail. There is no "fingerprint" channel pattern on the mantle *(Family Limacidae, most).*

- Go to 8

6c. The slug has a rounded back with no keel. Posterior end of tail is rounded in top view with a conspicuous posterior mucus pore. There is no "fingerprint" channel pattern on the mantle *(Family Arionidae).*

- Go to 9

7a. The keel extends from the tail all the way to the mantle. A horseshoe-shaped groove is found on the mantle. Mantle lacks "fingerprint" channel pattern *(Family Milacidae).*

- *Milax gagates* – greenhouse slug *(non-native)*

7b. The keel does not reach all the way to the back of the mantle. Mantle lacks a horseshoe-shaped groove. A "fingerprint" channel pattern is found on the mantle *(Family Limacidae).*

- Go to 8

8a. The animal's top surface is dark or pale, and may be speckled; adults are medium-sized (extended length 3.5 to 5 cm). Exudes milky mucus when irritated. Breathing hole on right side with a pale border.

● *Deroceras reticulatum* – gray fieldslug *(non-native)*

8b. The top surface is dark-colored and adults are small *(extended length 1.5 to 2.5 cm; often fast-crawling).* Mucus colorless even when irritated. Breathing pore on right side without a pale border.

● *Deroceras leave* – meadow slug

8c. The top surface is heavily mottled on front, with a mottled stripe or uniform color on the back part; adults are very large *(10 to 20 cm).*

● *Limax maximus* – giant gardenslug *(non-native)*

8d. Mantle and back with bands, not blotches; medium sized *(5 to 7 cm).*

● *Limax valentiana* – threeband gardenslug *(non-native)*

9a. The slug has orange color on its body.

● Go to 10

9b. The slug is without orange color on its body *(though it may be pale yellow).*

● Go to 11

10a. Orange is found on the side both above and below a dark side stripe. The mucus is orange *(to see the mucus color best, smear the slug against white paper).* Adults are up to 7cm.

● *Arion subfuscus* – dusky arion *(non-native)*

10b. Orange is found only *below* the dark side stripe, the sole of the foot is pale. The mucus is clear. Adults are up to 5cm.

● *Arion fasciatus* – orange-banded arion *(non-native)*

10c. There is a dark side stripe, but no orange on the side. The sole is orange or yellow, but the mucus is clear. Adults are up to 4cm.

● *Arion hortensis* complex – garden arion *(non natives);* this is a complex of three species – *Arion hortensis, Arion distinctus, and Arion owenii* - that are best told apart by dissection.

11a. There is a dark side stripe. The back is gray-yellow back and the sole nearly colorless. There are no pronounced bumps on the tail. Adults are up to 4cm.

● *Arion circumscriptus* – brown-banded arion *(non-native)*

11b. The side stripe is faint or absent. The back and sole are pale yellow. There are pronounced bumps on the tail when the animal is contracted. Adults are small, only up to 2cm.

● *Arion intermedius* – hedgehog arion *(non-native)*

Source: Carbondale, Illinois. Land snail identification and conservation workshop. June 30 - July 3, 2008).

Scientific Name

Bulimus gracil Hutton

Classification

Phylum	:	Mollusca
Class	:	Gastropoda
Sub Class	:	Orthogastropoda
Order	:	Pulmonata
Family	:	Subulinidae
Genus	:	Bulimus
Species	:	*bicolor*

Reference

Bulimus gracil, Hutton, 1834, *J. Asiat. Soc. Bengal, Calcutta*, Vol. 3: 84

Description

The shell is small (height 11mm) thin, somewhat translucent and glossy. It causes damages to potted plants and garden crops. It is wide spread in India, East Africa to Asia. They are commonly found in river bed.

Scientific Name

Cerastus moussonianus Petit

Classification

Phylum	:	Mollusca
Class	:	Gastropoda
Sub Class	:	Orthogastropoda
Order	:	Pulmonata
Family	:	Enidae
Genus	:	Cerastus
Species	:	*moussonianus*

Reference

Cerastus moussonianus, Petit, 1851, Jorn, Conchyl. Vol. 2; 266

Description

Externally the shell is traversed by very fine obliquely transpiral striae. The colour of the shell is whitish. General height is 18 mm and diameter is 10.5 mm. Their habitate is sandy bottom of the river. They are found in India in Ahmedabad, Mumbai, Ahmednagar, Broach, Khandala and Surat.

Scientific Name

Fastosarion brazier

Classification

Phylum	:	Mollusca
Class	:	Gastropoda
Sub Class	:	Orthogastropoda
Order	:	Pulmonata
Family	:	Subulinidae
Genus	:	Fastosariaon
Species	:	*brazier*

Description

This snail's shell is translucent and the soft mantle actually stretches up and over the outside of the shell.Thus, when the snail is sitting on a piece of old wood or rough tree bark, it is beautifully camouflaged. In this photo, the mantle has been drawn back to reveal the shell. This species is common in the Wet Tropics area.

Scientific Name

Gulella (Indoennea) bicor Hutton

Classification

Phylum	:	Mollusca
Class	:	Gastropoda
Sub Class	:	Orthogastropoda
Order	:	Pulmonata
Family	:	Enidae
Genus	:	Gulella
Species	:	*bicolor*

Synonyms

Pupa bicolor, Hutton 1834

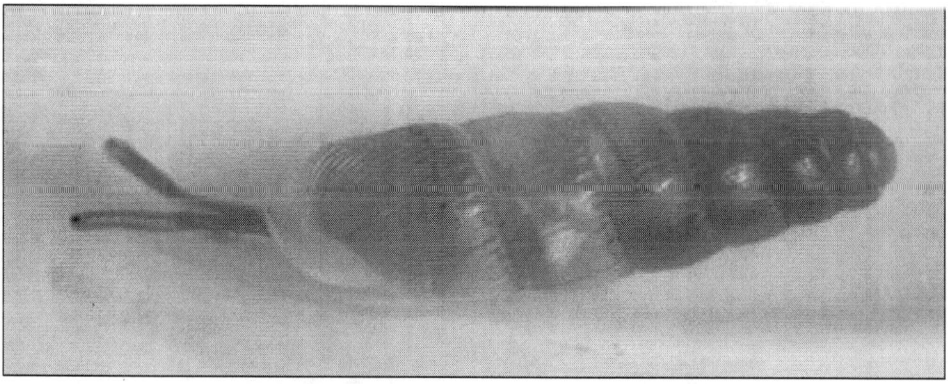

Reference

Gulella bicolor Hutton, Ray, 1923, *Rec.Ind. Mus.*; Vol 15: 390.

Description

Body whorl is compressed and bears funnel shaped depression on each side. The surface of the shell is smooth and glossy and white in colour. *Gulella bicolor*, whose shell grows to be about 7 mm long, is a predator of other equally small snails. In one study, starving captive individuals of *G. bicolor* did not eat several plant species given to them, suggesting that they are strict carnivores. They are available in sandy bottom of the river bed.

Scientific Name

Indrella ampulla (Benson, 1850)

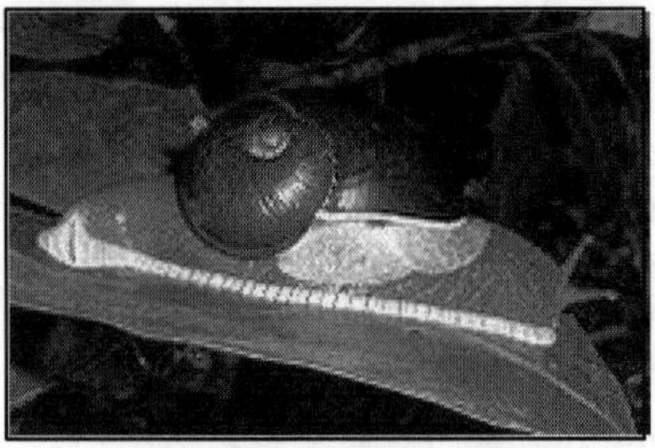

Classification

Phylum	:	Mollusca
Class	:	Gastropoda
Order	:	Pulmonata
Super Order	:	Stylommatophors
Family	:	Subulinidae
Genus	:	Indrella
Species	:	*ampulla*

Description

The shell is obliquely ovate and globose in shape and very thin. Half the thickness consists of epidermis, marked throughout with plicate line of growth, crossed by faint impressed spiral lines, and on the last whorl by shallow irregular furrows. The color of the shell is brownish olive, sometimes darker brown. Spire is small, convex and obtuse. The number of whorls is 3 and a half, rapidly increasing, the last much larger, rounded at the periphery and beneath. The aperture is very large, oblique, roundly oval, the same colour within as without, but smooth and glossy. The peristome is thin, membranaceous, columellar margin much curved inwards.

The width of the shell is 38-63 mm. The height is 30 mm. The width of the aperture is 32 mm and the the height of aperture is 30 mm. The color of soft parts has great color diversity, including white, cream, pale yellow, orange, red and black.

Scientific Name

Pupoides coenopictus Hutton

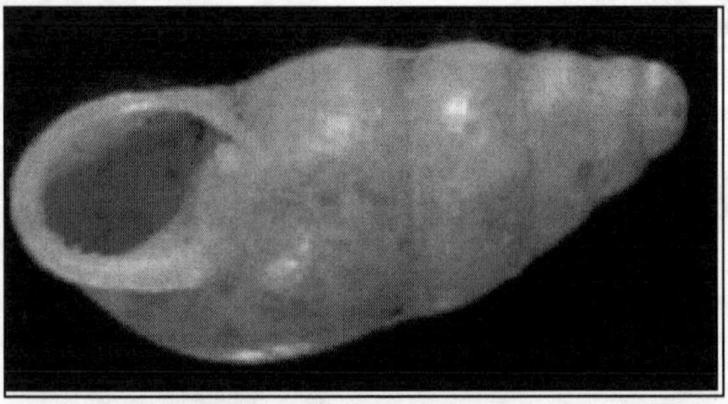

Classification

Phylum	:	Mollusca
Class	:	Gastropoda
Order	:	Pulmonata
Family	:	Pupillidae
Genus	:	Pupoides
Species	:	*coenopictus*

Reference

Pupoides coenopictus, Ray and Mukherjee, *1969, Rec. Zoo.Sur. India.;* Vol 61; 412.

Description

Body whorl is ventricose, while the remaining whorls are tapering to an obtuse apex. Aperture is ovato-lunate and is provided with parietal lamella. General height is 5.5 mm. It is found all over India and also in Afghanistan, Ceylon, and Burma.

Scientific Name

Rachis punctatus Anton

Classification

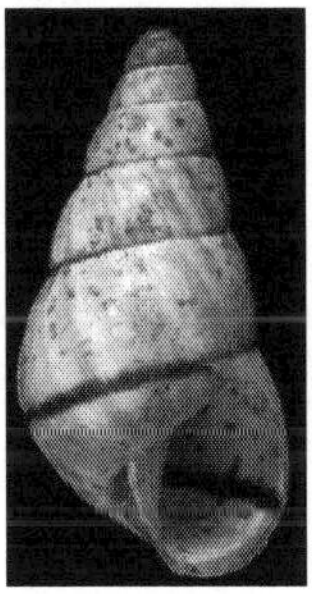

Phylum : Mollusca

Class : Gastropoda

Sub Class : Orthogastropoda

Order : Pulmonata

Family : Enidae

Genus : Rachis

Species : *punctatus*

Reference

Rachis punctatus, Satyamurtim 1960, *Bull. Mad. Govt. Mus;* Vol. 6(4): 74.

Description

The shell is pale brownish white in colour and marked with numerous irregularly scattered brownish spots. A characteristic single dark brownish spiral band is present slightly below the middle of the body whorl. Height is 9 mm. It is found in India, Africa, Ceylon, and Archpelago. They are very common in the sandy bottom of the entire river bed.

Scientific Name

Zootecus insularis Ehrenberg 1831

Classification

Phylum : Mollusca

Class : Gastropoda

Order : Pulmonata

Super Family : Stylommatophors

Family : Subulinidae

Genus : Zootecus

Species : *insularis*

Reference

Zootecus insularis, Ray and Mukherjee, 1969, Rec. Zool. Surv. India.; Vol. 61: 413.

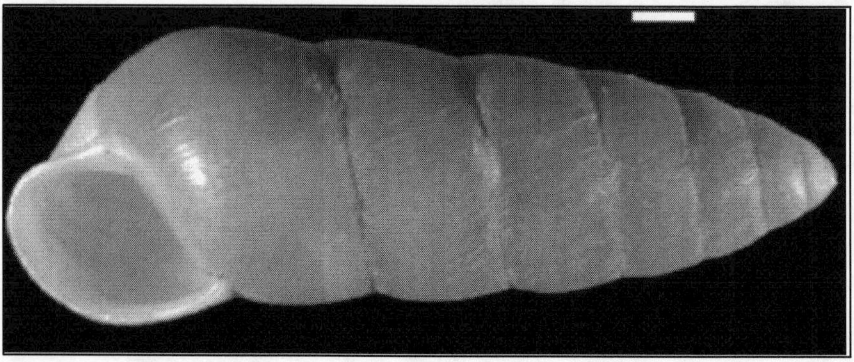

Description

The shell is somewhat transparent, horny, white in colour and with a glossy surface. The columellar margin of the aperture is short, hard and reflected. It is commonly found in sandy and muddy bottom of rivers.

Chapter 7
Commercial Use of Molluscs

Products Useful for Cottage and Small Scale Industries

Very little was known about uses of molluscs except for obtaining pearls, pearl dust and for using chank for puja.

Molluscs form one of the important fisheries in the Gulf of Kutch and area near by Sacred chank (*Xancus pyrum*), Ciwries (*Cyorea sp,*), Pearl Oyster (*Pinctada fucata*), window pane oyster (*Placenta placenta*, squid (*Loligo* sp.) and Cuttle fish (*Sepia* sp.) are known commercially important. But other molluscs or parts there also could be fruitfully used. Main use is to use them as food items.

Food

Along the world's miles of coastline, man has always had a readily available food source - high in protein and trace minerals, because of the many kinds of molluscs to be found there. Mussel and oyster beds, clam-flats and other abundant shellfish have always provided an easy source of food.

Tastes in molluscan food vary tremendously from one person to the next and from culture to culture; however, when it comes to a question of survival, most molluscs are edible. Some are considered delicacies such as oysters and escargot, while others such as the clams and mussels of freshwater ponds and streams are less likely to be consumed due to taste - but none-the-less are very edible! Terrestrial molluscs are also eaten.

Shell fish are recognized as nutritive sea food rich in essential food constituents. Some moluscs are known to have high vitamin and mineral content. They are very popular in Japan, Europ and America. In India molluscs are used as food normally by poorer classes living along the coastal belt or are utilized as exotic food dishes by star hotels. They have still to find acceptance as regular table dish by masses.

Majority of edible molluscs are distributed in three classes *viz.* gastropods, bivalvia, and cephalopoda. Proximate composition of molluscs is 9 – 20 per cent protein, 4 – 7 per cent of carbohydrate, 1 – 3 per cent of fat and 2 – 4 per cent of minerals. Squid meat contains 18 – 20 per cent protein said to be a little on higher side than many fishes. Squid are relished as food in Japan. Edible portion of squid is about 76 per cent of body weight, of which the mantle forms 42 – 43 per cent, fin 8 per cent and arms and head 24 – 25 per cent.

- ☆ Scallops (Family Pectinidae): There are many different kinds ("species") of scallops eaten in many parts of the world
- ☆ Oysters (Family Ostreidae): When eaten raw, are often thought to be an aphrodisiac. Oysters can get expensive, which is probably why there is a recipe known as Oysters Rockerfeller!"
- ☆ *NEVER* eat mussels or any other bivalve when you don't know if the area has a problem with this deadly kind of food poisoning!!! In a long-term survival situation, you could perhaps eat a very small nibble, then wait a day or two to see if you get ill, before going "whole quahog" over them.
- ☆ Whelks (Family Buccinidae)- VERY big in Japan, and the French pickle them for winter eating!
- ☆ Cockles (Family Cardiidae): (Europe & Malaysia) - as in Mary Molone, who cried "Cockles and mussels, alive, alive-o!" (shellfish are one of the few kinds of food that are kept alive right up to the moment of cooking: this is because they go bad very quickly, once dead!)
- ☆ Conchs (Family Strombidae): produce a large amount of meat, and are especially treasured in the Caribbean region, where the giant Pink (or Queen) Conch (*Strombus gigas* Linne) is farmed extensively. *Strombus gigas* in the Bahamas is heavily exploited for food, and along with two species of *Cassis* (Helmet shells), are often cultivated in Aquaculture farms.
- ☆ Top Shells (Family Trochidae: - Caribbean especially, where the West Indian Top Shell (*Cittarium pica* Linne, 1758). is considered a delicacy by some of the locals.
- ☆ Abalone (*Haliotis* spp): a well-known Tasty Morsel in many parts of the world. Some rare species can sell for up to $300 a pound ($760 per Kilo)!
- ☆ Periwinkles (Family Littorinidae): These, once again, are much loved by the French, who like their seafood, even when it is very small!
- ☆ Coquinas (*Donax* spp)(Caribbean): - Tiny clams, for chowder!
- ☆ Limpets (Families Patellidae, Fissurellidae, and Acmaeidae: Example: "Opihi" (*Cellana exarata* (Reeve)) lives on the rocks in Hawaii.
- ☆ Chitons (Class Polyplacophora): several families used for food). Example: Pacific Northwest Native Americans ate the Giant Pacific Chiton (*Cryptochiton stellari* (Middendorf)), which gets up to 300mm (12") long!
- ☆ Turban shells (Family Turbinidae:(Entire Indo-Pacific, but especially Japan, where, most of the time, if it comes from the sea, they will eat it!)

- ☆ Helmet shells (Family Cassidae: These can grow very large indeed: a 300 - 350mm (12-14") Horned helmet (*Cassis cornuta* Linne, 1758) could feed a whole family!
- ☆ Giant clams (Family Tridacnidae: Although they are rare and protected today, just imagine the amount of meat a fat, 1300mm (nearly 4 feet!) Giant Clam (*Tridacna gigas* (Linne)) could produce! The Bear Paw clam (*Hippopus hippopus* Linne) is still popular today in kitchens in the Philippines.
- ☆ *Busycon carica*: is an important ingredient in the Italian dish of "scungilli marinara", and is commonly referred to as the Bulot shell.

  Only a few molluscs are actually poisonous. A wide variety of molluscs end up in cooking pots around the world every day.
- ☆ Pen shells (Family Pinnidae: They reportedly taste similar to scallops.

(*Source*: Internet)

- ☆ Squid & octopus: Particularly popular in Japan (where squid are called "Calamari"), but they are catching on in parts of the Western World. In Newfoundland (Canada), they are used by the ton as bait for several kinds of fish.(Note: These are cephalopods - a totally different Class of molluscs from the gastropods or bivalves. They are the most developed molluscs, and some are regarded as the most intelligent invertebrates (animals without a backbone) in the world!) Mantle with fins and skin was cooked in boiling water for about 15 minutes. Skin and fins were then removed after which desired sized rings were prepared by cutting with sterilized scissors. Rings were sundried. Such squid rings can be used by frying.

  Squids can be frozen. Frozen squid meat is finding good export in international market, particularly in Japan

Edible Oyster Meat

Edible oyster meat is also relished very much abroad. Hotels, particularly in coastal metropolitan cities in India are also serving a number of oyster meat dishes.

Small Scale Industries

Molluscs shells contain about 33 40 per cent calcium and approximate 90 – 98 per cent calcium carbonate. They would thus be useful raw material in any calcium/ calcium carbonate based industries. They would also be useful in industries using calcium carbonate as filler. It is possible to prepare calcium carbide, calcium cynamide and derivatives of calcium carbide.

However it is important to assess the annual availability of required types of shells as a valuable resources and also arrive at estimates of optimum quantum, utilizable in given areas or else there could be a problem of depletion of stocks resulting in wiping out of natural resources of particular types of mollusc shelled species.

Considering above very important aspect of conservation and keeping in view the need of "Utilization" and not exploitation of molluscs shell resources first of all survey has to be conducted. Some of the cottage industries, without affecting natural stock, can be established. They are as under:

Shell Grit

Small pieces of dead shells found at coastal areas may be collected and grind in small pieces. Mix varieties can be used. This type of shell grits can be utilized in preparing poultry feed.

Interior Decoration Items

Beautiful interior decoration items, jewelry items as well as toys and curies can be prepared out of different molluscs shells. Bengals with beautiful carvings, rings, pendants, necklaces, buttons and many other items of interest and commercial value can be prepared out of chanks. Chank shell supports an important industry preparing these items of the Tirunenelli and Ranathpuram coast of South India as well as West Bengal and Bangladesh. They get considerable percentage of chank requirements from the Gulf of Kutch.

Molluscs in Medicine

Shell amulets were once thought to ward off ill health, infertility or bad luck. Shells have also been ground up for use in portions and for various medicinal uses throughout history. Today the shell, its living flesh and by products are being studied and used in many areas of medicine. Some examples:

- ✰ The deadly venoms of some Cone Shells (Conidae) are today being used to help victims of strokes and heart disease, and to produce a revolutionary new drug for chronic pain control (Ziconotide - still awaiting FDA approval)
- ✰ An extract from the hard clam or "Quahog" (*Mercenaria mercenaria* L.) is a strong growth inhibitor of cancers in mice. It is called mercenine, after the clam's scientific name.
- ✰ Ground and processed oyster shells are used as a calcium supplements both for humans and animals.
- ✰ Paolin, a drug made from abalone juice, is an effective inhibitor of penicillin-resistant strains of bacteria, such as *Staphylococcus aureus, Streptococcus pyogenes, Salmonella typhus* and parathyphoid A & B bacteria.
- ✰ Oyster juice has been found to have anti-viral properties, and may be made into a drug eventually.
- ✰ The threads that some mussels (Mytilidae) use to attach themselves to rocks, piers, and other hard surfaces are being tested as possible glue in surgery. Note: These are called "byssal" threads, from the Latin word byssus, which means "fine linen", which is silky, like the fine threads of many molluscs. Quite often, a Latin or Greek word borrowed by science in this fashion.)
- ✰ The cement of the Carrier Shells (Xenophoridae) is being studied for use as a possible cement for bone fractures.Note: The Carrier shells are the camouflage experts of the mollusc world: they attach all kinds of objects - shells, rocks, pieces of coral, sponges, bottle caps to their shells, so they

look like a little pile of trash on the bottom of the sea - a great way to avoid being eaten!

- ☆ Powdered mother of pearl is sold in the markets of the Peruvian Andes to promote healing of wounds (unproven). *Note*: Mother of pearl is the beautifully iridescent interior of some bivalves, which the mollusc also uses to surround small, irritating objects which get inside its mantle, or outer body - forming a pearl!)
- ☆ Traditional medicine in Vietnam has a wide variety of uses for shells: powdered oyster shell is taken to treat acid indigestion, fatigue and to stop hemorrhage. It is also sprinkled over open wounds and boils. Cuttlefish bones are used as a remedy for rickets, a healing agent in the treatment of gastro-intestinal troubles, a local anti-hemorrhagic (*i.e.*, it stops internal bleeding), and as an antiseptic is cases of inflammation of the middle ear. The flat shell of the Abalone, with its iridescent inside, is powdered and taken orally to improve vision, to remove keratoses (cataracts), and to improve such conditions as hemeralopia (where you can see at night well, but hardly at all in the daytime! Powdered pearls from oysters are used as a topical eye medicine (*i.e.*, you put it right on your eyes!!), and it has been scientifically proven to have some anti-inflammatory effects on a painful condition called conjunctivitis, where the surface of the eye becomes red and sore.

Ayurvedic Medicine

Some molluscs are used as medicine since long. Charka, the great ayurvedic medicine expert has written in detail about the use of various molluscs shells and animal parts in ayurvedic treatments.

Flesh of some species of univalves as Lambis is said to have narcotic effect. Chank powder and cowrie powder prepared in specific ways are used in some skin ailments eye diseases, asthama, rickets, cough and colds etc. The apple snails also have some medicinal properties. The ark shell meat is often used as medicine in cases of pregnancy. "Surma" an ayurvidic medicine used for some eye treatment is prepared from the pearls of window pane oyster. Cuttle bone powder mixed with oil is used for some ear troubles in coastal areas. Local inhabitants of Okhamandal and surrounding areas consider dried chank flesh to very useful in certain specific ailments.

Coastline could form an important source of food, raw material for village industries, art-craft industry, indigenous medicine raw material and other things, providing potential for employment to large number of people in the village near coastlines.

Trade Goods

Shell currency has been around for over 4,000 years and was, in its heyday, the most widely used currency in the world. Even today, there still exist minor currencies based on certain shells.

Some examples of shells' uses in trade are:

- Cowrie shells (*Cypraea annulus* L., and *C. moneta* L.), collected loose in bags or strung into strands, were the earliest forms of currency used in many countries. The Chinese, so far as we know, were the first people to use cowries as currency. Here, cowries have been found in prehistoric Stone Age sites. Examples of other country's native money-strands are the diwara in New Guinea, rongo in the Melanesian islands and sapisapi in Africa. The image of the cowrie as a type of currency was so strong that the first oval metal coin minted in the Greek colony of Lydia around 670 B.C. was modeled after that shell. By the eighteenth century, approximately 400 million cowries were being traded per year mostly for the purchase of black slaves. By the middle of the nineteenth century, it could take up to 100,000 cowries just to buy a young wife. Inflation, it seems, was the main demise of the cowrie currency.
- Hard clamshells and whelks were the shells used to make the North American Indian wampum. Eastern Indians also used the tusk shell *Dentalium pretiosum* Sowerby (collected on Vancouver Is.!! (Canada)) Sowerby, as a trade shell. Wampum continued to be used as money through the first half of the eighteenth century when it finally died out due to counterfeiting and mass production.
- Beads and other ornaments made of"Spiny Oyster" shells (*Spondylus princeps* (Broderip) and the Panamanian Pearly oyster *Pinctada mazatlantica* (Hanley) were traded all over the Andean region. The Chumash Indians of California also make shell beads from the purple olive (*Olivella*) shell that they use as money.

Tools

From prehistoric times, man has used shells for tools. This practice has been born out by archaeological findings in ancient sites and still carries on even today. Some examples of these shell tools are:

- Household dishes, cooking pots and utensils: cutlery, scoops, spatulas, etc. were often made from bivalves and larger gastropods such as the bailer (*Melo* (a Volute), whelk *Nautilus* and turban (Turbinidae) shells
- Food pounders were made from the crystalline stomach style (which the animal uses to help digest its food) " giant clam" (*Tridachna gigas* L.), in the South Pacific.
- Storage containers for such things as perfumes, ointments and medicines were made from some of the larger bivalves and univalves such as the Nautilus.
- Oil lamps made from shells are a frequent find throughout the Middle East. There are examples of these made from bear paw shells (*Hippopus hippopus* Linne) and the spider conch (*Lambis* spp). They work by holding oil while the wick floats on the surface.

- Fishing gear: fish lures, octopus lures, hooks and sinkers were made from abalone, pearl shell and cowries.
- Tweezers, tongs and claspers were made from bivalves (the two halves making them ideal for this!).
- Farming Tools: Shovels, plow blades and hoes for tilling the soil were made from hard, sharpened shell pieces.
- Building tools: designed to split and smooth many building and thatching materials such as palm fronds and bamboo canes.
- Blades and scrapers for cutting and skinning hides were made from shells such as the ark shell (family *Arcidae*)
- Adze, knife and axe blades were made from shells with sharpened edges.
- Drills, chisels, scrapers, sanders, etc. were made from various shells such as the Red Helmet shell (*Cassis rufa* L.) and were used in such trades as woodworking, farming and tool making.
- Weapons such as spearheads and gouges were made from sharpened and shaped pieces of hard shells such as the Queen conch (*Strombus gigas* L.).
- Bailing buckets made from "bailer" shells (*Melo*) are still in use by native fisherman in the South Pacific and Australia today - to bail out their boats!

Architecture

- Leonardo da Vinci drafted the first spiral staircase plans (which are still used by architects today!) from studying the simple snail shell with its interior whorls.
- The spiral tops of Grecian Columns were designed after a nautilus shell cut in half.
- During the Renaissance, architects copied shell shapes for design in niches, facades, tombs and pedestals.
- Frank Lloyd Wright designed the Guggenheim Museum around the Japanese Miracle Shell (*Thatcheria mirabilis* (Angass)).
- The Sydney Opera House was designed to look like a giant Cock's Comb Oyster (*Lopha cristagalli* (Linne)). Go to the John Uzton Page for 2 great pictures of this building (the second one is best!)
- The Mayans of Mexico carved bivalves and conch shells into the walls of their public buildings and temples.

Decorations

- Man has been using shells to decorate his dwellings and public meeting places since before the dawn of history.
- Ancient Greeks collected shells to decorate their gardens and fishponds. At the height of the Rococo era, real shells for decoration became vogue. Shells were especially used to make, little houses and grottos in the gardens and parks of great chateaux and houses in France and England.

Religion

- ☆ Shells have played a central role in religion from prehistoric times on. The presence of shells in prehistoric burial places indicates that their symbolic power was believed to continue beyond life. Shells in some cultures even today are used as amulets, good luck charms, and as symbols for love, fertility and life eternal.

Some examples of some these religious practices are:

India: Hindus: The god Vishnu holds his staff crowned with a very rare left-handed *Turbinella* ("Chank") shell The Hindu, when praying, often clasps a sacred chank or other venerated object in his hands, believing that it will help his or her petitions be heard. Priests also use it for holding sacred oils.

- ☆ *Africa*: Shells fetishes were often used in worship. Ceremonial garbs are many times decorated with shells and were used in some religious ceremonies. North American Indians also made fetishes of shells. The Canadian Ojibwa tribe maintained a Grand Medicine Society in which the sacred emblem was a shell.
- ☆ *Asia: Buddhists*–The Chank or *Turbinella* also plays a significant role in their ritual music and ceremonies, and figures into Buddhist iconography.
- ☆ *Spain*: The home of the shrine of Santiago(St. James). St. James's badge is the Giant European scallop shell (*Pecten jacobaeus* Linne). Pilgrims to this shrine purchased the simple but exquisitely sculptured scallop shells and wore them as a sign of their pilgrimage to the shrine. This scallop also appears in many paintings and statues of this saint throughout Europe.
- ☆ *Egypt, China* and other cultures used the cowry in connection with their burials and other religious ceremonies.Sierra Leone: Cannibals during the nineteenth century used cowry shells in part of their ceremonial rituals.
- ☆ *Pre-Columbian South and Central America*: Archaeological sites have produced shell trumpets that may have played a role in religious ceremonies. In the Andes region, a Thorny Oyster (*Spondylus princeps* (Broderip)), and the giant E. Pacific conch (*Strombus galeatus* Linne) as well as *Pinctada atlanticus* (a Winged Oyster) all had important religious significance. The Aztecs of Mexico also used shells in their religion: Tlaloc, the rain god, is depicted as emerging from a conch (*Strombus* spp) shell. They also used conch and horse conch (*Pleuroploca*) shell trumpets.
- ☆ *Minoan Crete*: Shell trumpets were used in religious ceremonies.
- ☆ *Christianity*: Many churches had or still have baptismal fonts made of *Tridacna gigas* Linne) (the famous "Giant Clam"!) or are designed in their likeness. They are thought to be a symbol of birth.

  (Avril Bourquin, Ross Mayhew and many contributors 1999. (Man and Mollusc Uses of Shell-Bearing Molluscs - Past, Present & Future)

APPENDICES

APPENDIX-I

List of Edible Molluscs of the Gulf of Kutch and the Northern-Western of Saurashtra Gujarat, india

Sl.No.	Family	Scientific Name	Common Name	Collection Area
CLASS: AMPHINEURA				
1.		*Ischnochiton Sp.*	Coat-of-mail shells	Okha, Cgushapir.
CLASS: GASTROPODA				
2.	PATELLIDAE	*Cellana radiate*	Limpets	Okha, Hanuman-dandi
3.	TROCHIDAE	*Trochus stellatus*	Top shells or botton shells	Okha, Adatra, Poshetra, Hanumandandi, Chushapir, Baide, Nora
4.	TROCHIDAE	*T. radiates*	Botton shell	-Do-
5.	TROCHIDAE	*Umborrnium vestriarium*	Top shell	-Do-
6.	TURBINIDAE	*Turbo intercostalis*	Turban shell	Balapur, Okha, Adatra
7.	TURBINIDAE	*T. coronatus*	Turban shell	Poshitra, Dwarka, Boria, Pashu, Nora, Chusnapir, Baida
8.		*Astrea semicostata*	Turban shell	Okha, Poshitra,Hanuman-dandi, Pashu.
9.	NERITIDAE	*Nerita albicilla*	Nerites	Okha, Adatra, Poshitra, Hanuman-dandi.
10.	NERITIDAE	*N. dombeyi*	Nerites	Okha, Adatra, Poshitra, Hanuman-dandi.
11.	POTAMIDIDAE	*Telescopium telescopium*	Horn shell	Sikka, Baide, Balapur Bay.
12.	STROMBIDAE	*Strombus urecus*	Wing shell	Hanuman Dandi
13.	NATICIDAE	*Natica tigrina*	Shankula (Gujarati local name)	Okha, Hanuman Dandi, Baida
14.	NATICIDAE	*N. kambarcki*	Shankhula	Hanuman Dandi
15.	MURICIDAE	*Murex virgineus*	Murices	Okha, Adatra, Poshetra, Boria
16.	MURICIDAE	*M. adustus*	Murices	Baida, Nora, Pashu, Hanuman dandi
17.	MURICIDAE	*Thais rugosa*	Purple	Okha, Poshetra, Off Pindhara

Contd...

Appendix-I–Contd...

Sl.No.	Family	Scientific Name	Common Name	Collection Area
18.	MURICIDAE	*T. rudolphi*	Purple	Okha, Poshetra, Off Pindhara
19.	MURICIDAE	*T. carinifera*	Purple	Okha, Poshetra, Off Pindhara
20.	VASIDAE	*Xancus pyrum*	Scred chank	Pkha, Aramda, Baida, Pashu, Nora.
21.	OLIVIDAE	*Oliva gibbosa*	Olives	Okha, Hanuman dandi
22.	OLIVIDAE	*O. nebulosa*	Olives	Okha, Hanuman dandi
23.	BUCCINIDAE	*Babylonia spirata*	Whelks	Hanuman dandi, Balapur,Of Dwarka.
CLASS: PELECYPODA (BIVALVIA)				
24.	ISTREIDAE	*Crassostrea gryphoides*	Edible Oyster	Dwarka, Armada, Hanuman dandi, Harshad creek, Muddy creeks of Kutch, Porbundar
25.	ISTREIDAE	*C. discoides*	Edible Oyster	Dwarka, Armada, Hanuman dandi, Harshad creek, Muddy creeks of Kutch, Porbundar
26.	ISTREIDAE	*Ostrea madransis*	Edible Oyster	Dwarka, Armada, Hanuman dandi, Harshad creek, Muddy creeks of Kutch, Porbundar
27.	ISTREIDAE	*O. folium*	Edible Oyster	Dwarka, Armada, Hanuman dandi, Harshad creek, Muddy creeks of Kutch, Porbundar
28.	ISTREIDAE	*C. cucullata*	Edible Oyster	Dwarka, Armada, Hanuman dandi, Harshad creek, Muddy creeks of Kutch, Porbundar
29.	PTERIIDAE	*Pinctada fucata*	Pearl Oyster	Pirotan, Movada, Narara, Ajad, Sikka
30.	PINNIDAE	*Pinna vexillum*	Pen shell or Feather shell	Poshitra, Sikka, Hanuman dandi, Baida Pirotan
31.	PINNIDAE	*P. bicolor*	Pen shell or Feather shell	Poshitra, Sikka, Hanuman dandi, Baida Pirotan
32.	PINNIDAE	*P. atroparpurea*	Pen shell or Feather shell	Poshitra, Sikka, Hanuman dandi, Baida Pirotan
33.	PINNIDAE	*P. nigra*	Pen shell or Feather shell	Poshitra, Sikka, Hanuman dandi, Baida Pirotan

Contd...

Appendix-I–Contd...

Sl.No.	Family	Scientific Name	Common Name	Collection Area
34.	VENERIDAE	*Paphia textile*	Clam	Baida Poshitra, Pirotan
35.	VENERIDAE	*P. malabarica*	Clam	Baida Poshitra, Pirotan
36.	VENERIDAE	*Meretrix,eretrix*	Clam	Baida Poshitra, Pirotan
37.	VENERIDAE	*M. casta*	Clam	Veraval.
38.	VENERIDAE	*Gafrarium tumidum*	Clam	Pirotan
39.	OSTREIDAE	*Placenta placenta*	Window pane oyster	Balapur Bay, Pindhara Bay
40.	MESODESMATIDAE	*Mesodesma glabratum*	Small clam	Pirotan
41.	DONACIDAE	*Donax cuneatus*	Wedge shell	Hanuman dandi
42.	SOLENIDAE	*Solen lamarckii*	Razor shell	Balapur Bay
43.		*S. truncates*	Razor shell	Balapur Bay
44.	ARCIDAE	*Arca granosa*	Ark shell	Pirotan
45.	MYTILIDAE	*Mytilus viridis*	Green Muscle	Pirotan
46.		*Modiolus metcalfei*	Weaving muscle	Pirotan.

Source: Sarvaiya, R. T. 1983.

APPENDIX-II
List of Freshwater Molluscs

Sl.No.	Scientific Name	Name of Scientist
1.	*Cyclophorus (Litosstylus) involvulus*	Muller
2.	*C. zurantiacus*	Schumbacher
3.	*Vivipara bemgalensis*	Lamarch
4.	*V. dissimillies*	Muller
5.	*Pila globose*	swainson
6.	*Birthynia stenothyroides*	Dohn
7.	*Sulcospira hugely*	
8.	*Faunus ater*	(Linne)
9.	*Pauludomus (Stomatodon) stomatodon*	(Benson)
10.	*Melania (Plotia) scabra*	Muller
11.	*M. tuberculata*	Muller
12.	*Lymnaea luteola*	Lamarck
13.	*L. acuminate*	Lamarck
14.	*L. pignguis*	Dhorn
15.	*L. auricularia*	Draparnaud
16.	*Planorbis (indoplanorbis) exustus*	Deshayes
17.	*Anisus (Gyraulus) convexiusculus*	Hutton
18.	*Opeas gracile*	Hutton
19.	*Zootecus insularis*	Ehrenberg
20.	*Z. chion*	Pfeiffer
21.	*Macrochlamys insfasta*	Blanfford
22.	*M. pedina*	Benson
23.	*Cryptozona (Xestina) belangeri Var. bombayna*	Pfeiffer
24.	*Ariphanta laevipes*	Muller
25.	*A. Bajadera*	Pfeiffer
26.	*Planispira proxima*	Ferussac
27.	*Parreysia corrugate Var. negpoorensis*	Lea
28.	*Lamellidens marginalis*	Lamarck
29.	*L. corrianus*	Lea
30.	*Corbicula regularis*	Prime
31.	*Cyclophorous (glossostylus) indicus*	Deshayes
32.	*Cyclotopsis semistriata*	Sowerby
33.	*Digoniostoma pulchella*	Benson
34.	*Alocimna orcula Var. product*	Nevill

Contd...

Appendix-II–Contd...

Sl.No.	*Scientific Name*	*Name of Scientist*
35.	*Paludomus obese*	Phillippi
36.	*Rachis pinctatus*	Anton
37.	*Glessula Notigena*	Benson
38.	*G. ceylanica*	Pfieffer
39.	*G. (Rilshetia) dilkarngense*	Godwin-Austen
40.	*Sitala denselirata*	Preston
41.	*Kaliella bullula*	Huton
42.	*Macrochlamys tenuicola*	Adam
43.	*M.platychlamys*	Adem
44.	*Xesta (Fretum) semirugata*	Beck
45.	*Ariophanta bistrialis*	Beck
46.	*Planispira (Trachia) crassicstata*	Benson
47.	*Eulota scalpurita*	Benson
48.	*Indonaia coerulea*	Lea
49.	*Corbicula Striatella*	Deshayes
50.	*C. peninsularis*	Prashad
51.	*Peltodoris rubescens*	Bergh
52.	*Platydoris pulchra*	Elliot
53.	*Achatina fulica*	Bowdich
54.	*Viviparus bengalensis*	Lamarck
55.	*V. variatus*	Frauenfeld
56.	*Digoniostoma pulchella*	Benson
57.	*Melanoides tuberculatus*	Muller
58.	*M. (Terebia) lineatus*	Grey
59.	*M. (Plotia) scbra*	Muller
60.	*Indoplanorbis exustus*	Deshayes
61.	*Lamellidens consobrinus*	Lea
62.	*Lymnaea(pseudosuccinea) luteola*	Lamarck
63.	*L. luteola*	Lamarck
64.	*Gyraulus convexiusculus*	Hutton
65.	*Planispira asperella*	Pfeiffer
66.	*Eurychlamys platychlamys*	Blanford
67.	*Indonaia caerulea*	Lea
68.	*Corbicula striatella*	Deshayes

Source: Dholakia A. D., 2004.

APPENDIX-III

Locality of Molluscs at Various Centers of Saurashtra (Gujarat)

Sl.No.	Species	1	2	3	4	5	6	7	8	9	10	11	12	13	14	15	16	17
1.	*Chiton* sp.	P	P			P	P		P				P					
2.	*Celana radiate* (Born)	P								P						P	P	
3.	*Trochus radiates* Gmelin	P			P				P	P				P	P	P	P	
4.	*Trochus* sp.	P	P				P	P		P			P		P	P	P	
5.	*Astrea semicostata* (Kiener)	P			P	P	P			P				P		P	P	
6.	*Nerita domheyi* R'wcluz	P	P		P	P	P			P					P	P	P	
7.	*N. albicilla* Linn	P			P	P			P	P							P	
8.	*Turbo intercostalis* Menke	P	P		P	P	P	P	P	P	P	P	P	P	P		p	
9.	*T. coronatus* Gmelin	P	P		P		P	P	P	P	P	P		P		P	P	
10.	*Cerithidez fluviatilis* (Poti'ea & Michaud)	P	P	P					P	P				P	P	P	P	
11.	*Telescopium telescopium* Linn							P	P	P					P	P	P	p
12.	*Onchidium verruculatum* Cuvier	P	P	P		P	P	P	P	P		P	P	P	P	P	p	
13.	*Thais rugosa* Born	P	P		P	P	P		p	P	P				P	P	P	
14.	*T. eudolphi* Lamarck	P				P				P							P	
15.	*Thais* sp.						P	P	P	P					P	P	P	p
16.	*Murex virgineus* Lamarck	P	P		P	P	P	P		P	P			P	P	P	P	P
17.	*M. adustus* Lamarck	P			P	P				P				P	P		P	
18.	*Murex* sp.	P														P	P	
19.	*Hemifusus* sp.							P	P	P							P	
20.	*Babylonia spirata* (Linn)	P			P	P			P	P						P		
21.	*Aplysia* sp.	P	P				P			P							P	
22.	*Cypraea* sp.	P	P		P	P	P	P	P	P				P	P	P	P	
23.	*Oiiva gibbosa* (Born)	P								p								

Contd...

Appendix-III–Contd...

Sl.No.	Species	1	2	3	4	5	6	7	8	9	10	11	12	13	14	15	16	17
24.	*Oliva* Sp.	P	P		P					P							P	
25.	*Conus* sp.	P	P		P	P	P			P				P		P	P	
26.	*Tibia curts* Sowerby						P	P										
27.	*Xancus pyrum* (Linn.)	P		P	P		P							P	P	P	P	
28.	*Natics tigrina* (Roding)	P					P			P						P	P	
29.	*Nassa* sp.	P	P		P	P	P			P				P	P	P	P	
30.	*Turritela* sp.	P	P		P	P	P	P		P	P			P	P	P	P	
31.	*Bursa granularis* (Roding)	P			P					p								
32.	*Crasostrea gryphoides* (Newtona and Smith)			P		P	P	P	P						P	P	P	
33.	*C, cucullata*		P		P	P	P	P		P					P		P	
34.	*Placenta placenta* (Linn'e)						P		P						P	P	P	
35.	*Pinna bicolor* Gmelin							p										
36.	*P. atropurpurea* (Sowerby)						P	P								P		
37.	*Pinna* sp.							P	P						P	P	P	
38.	*Solen* sp.								p									
39.	*Cardium* sp.				P			P		P				P		P	P	P
40.	*Pitar erycina* (Linn'e)		P		P		P	P		P						P	P	
41.	*Pitar* sp.		P		P	P	P	P	P	P					P	P	P	
42.	*Lithophaga* sp.		P		P		P								P	P	P	P
43.	*Area* sp.		P		P					P					P	P	P	
44.	*Cuttle bone*	P	P		P	P	P			P						P	P	
45.	*Octopus* sp.	P	P				P	P		P					P	P	P	

Source: Sarviya R.T., 1983 and Dholakia A.D., 2004. IS: Island.

Center of availability: 1: Okha; 2: Adatra; 3: Aramda; 4: Mithapur; 5: Dwarka; 6: Poshetra; 7: Sikka; 8: Balapur; 9: Hanuman Dandi; 10: Boria Island; 11: Pagga Is; 12: Chusnapir Is; 13: Pashu Is; 14: Nora Is; 15: Baida Is; 16: Ajad Is; 17: Dhani Is.

APPENDIX -IV
Identification Details of Families in Bivalvia with Figure
(As per key to Families)

Sl.No.	Name of Family	Figure No.	Page No.
1.	Isognomonidae	A	12
2.	Malleidae	A	12
3.	Pteriidae	A	12
4.	Limidae	B	13
5.	Pectinidae	B	13
6.	Spondylidae	B	13
7.	Anomiidae	C	14
8.	Placunidae	C	14
9.	Plicatulidae	C	14
10.	Tridacnidae	D	14
11.	Gryphaeidae	E	14
12.	Ostreidae	E	14
13.	Pholadidae	F	16
14.	Teredinidae	F	16
15.	Mytilidae	G	17
16.	Pinnidae	G	17
17.	Arcidae	H	17
18.	Cucullaeidae	H	17
19.	Glycymerididae	H	17
20.	Noetiidae	H	17
21.	Laternulidae	I	18
22.	Peneplomatidae	I	18
23.	Thracugae	I	18
24.	Crassatellidae	J	20
25.	Mactridae	J	20
26.	Mesodesmatidae	J	20
27.	Semelidae	J	20
28.	Chamidae	K	20
29.	Solecurtidae	L	21
30.	Solenidae	L	21
31.	Lucinidae	M	21
32.	Thyasiridae	M	21
33.	Corbiculidae	N	22

Contd...

Appendix-III–Contd...

Sl.No.	*Name of Family*	*Figure No.*	*Page No.*
34.	Glauconomidae	N	22
35.	*Petricolidae	N	22
36.	Veneridae	N	22
37.	Donacidae	O	23
38.	Psammobiidae	O	23
39.	Tellinidae	O	23
40.	Cardiidae	P	24
41.	Carditidae	P	24
42.	Hemidonacidae	P	24
43.	Fimbriidae	Q	25
44.	*Ungulinidae	Q	25

APPENDIX-V
Glossary of Technical Terms

1. **Accessory plate**—calcareous and periostracal structure covering the soft parts in the Pholadidae, in addition to the shell valves.
2. **Adductor muscle**—muscle connecting the 2 valves of a shell, tending to draw them together.
3. **Apophysis**—finger-like shelly structure to which the foot muscles are attached in the Pholadidae and Teredinidae.
4. **Branchial**—pertaining to the gills.
5. **Branchial lamella**—(see gill).
6. **Byssus**—clump of horny threads spun by the foot, by which a bivalve can anchor to a hard substrate.
7. **Cardinal area**—surface of the shell extending between umbo and hinge margin.
8. **Cardinal tooth**—(see tooth).
9. **Chomata**—marginal crenulations in Ostreidae and Gryphaeidae, occurring all around the inner side of valves or only near the hinge, composed of small tubercles or ridgelets on the right valve, and corresponding pits on the left valve.
10. **Commissure**—line of junction of the valves.
11. **Concentric**—parallel to lines of growth.
12. **Cruciform muscles**—crossed muscles connecting valves and serving to retract the siphons, leaving 2 small scars near the posteroventral end of pallial line in some bivalves (*e.g.*, Tellinidae).
13. **Ctenidial axis**—(see gill).
14. **Ctenolium**—a row of small teeth on lower side of byssal notch in some Pectinidae.
15. **Demibranch**—(see gill).
16. **Denticle**—a small tooth.
17. **Ear**—lateral expansion of the dorsal part of a shell.
18. **Equilateral**—the condition of a valve when growth on either side of umbo is symmetrical.
19. **Equivalve**—the condition of a shell when valves are of the same shape and size.
20. **Escutcheon**—differenciated area extending along dorsal margin of valves, behind the umbones.
21. **Eulamellibranchiate type**—gill demibranchs composed of 2 lamellae. Branchial filaments and lamellae always connected by tissular junctions (*e.g.*, Veneridae).
22. **Filibranchiate type**—gill demibranchs composed of 2 lamellae. In addition to the ciliary junctions between branchial filaments, anastomosed tissular junctions may unite lamellae of each demibranch (*e.g.*, Mytilidae, Pectinidae).
23. **Foot**—mobile and extensible muscular organ, used for locomotion or for attachment to substrate by means of byssal threads.
24. **Gape**—opening or gap remaining between margins of valves, when shell is closed.
25. **Gill**—respiratory organ generally composed of 2 thin leaf-like structures (demibranches. suspended to a dorsal axis (ctenidial axis); each demibranch may be either simple or bent back upon itself and then formed of 2 sheets (branchial lamellae). A lamella is constituted of many ciliated filaments parallel to each other and interconnected by more or less

complex junctions. Four main types of gill structures are currently recognized among bivalves: the protobranchiate, filibranchiate, eulamellibranchiate, and septibranchiate types (see these terms).

26. **Growth marks**—(see sculpture).
27. **Hinge**—structures in the dorsal region of the shell, along which the valves meet, and that function in the opening and closing of the shell.
28. **Hinge line**—shell margin adjacent to the hinge.
29. **Hinge plate**—infolding of dorsal shell margin bearing hinge teeth and sockets, and lying in each valve in a plane parallel to that of junction of valves.
30. **Imbricate**—overlapping like tiles or shingles on a roof.
31. **Inequilateral**—the condition of a valve when growth on either side of umbo is a ssymmetrical.
32. **Inequivalve**—the condition of a shell when valves are not alike in shape or size.
33. **Keel**—a prominent, angular ridge.
34. **Lamellate**—with thin, flattened plates.
35. **Lateral tooth**—(see tooth).
36. **Lenticular**—shaped like a biconvex lens.
37. **Ligament**—horny, elastic structure joining the 2 valves dorsally.
38. **Ligamental area**—part of cardinal area occupied by the ligament.
39. **Lunule**—differentiated area extending along dorsal margin of valves, just in front of umbones.
40. **Mantle**—fleshy sheet surrounding vital organs and composed of 2 lobes, one lining and secreting each valve.
41. **Muscle scar**—impression marking the place of attachment of a muscle inside the shell.
42. **Nacreous**—pearly, often with multi-coloured hues, as in mother-of-pearl.
43. **Nymph**—narrow plateform extending behind umbo along dorsal margin, to which the external ligament is attached.
44. **Opisthogyrate**—the condition of a shell when umbones are directed posteriorly.
45. **Orbicular**—disk-shaped, nearly circular. 126 Bivalves
46. **Orthogyrate**—the condition of a shell when umbones are perpendicular to the hinge line (directed neither anteriorly nor posteriorly).
47. **Pallet**—small paddle-shaped or feather-like calcareous and periostracal structure, a pair of which closes the burrow opening when siphons are retracted in the Teredinidae.
48. **Pallial**—pertaining to the mantle.
49. **Pallial line**—a line near internal margin of valve, marking the site of attachment of the mantle edge.
50. **Pallial sinus**—posterior indentation of pallial line, marking the site of attachment of muscles allowing siphons to retract within the shell.
51. **Pedal**—pertaining to the foot.
52. **Periostracum**—layer of horny material covering the shell.
53. **Plicate**—folded or ridged.
54. **Porcelaneous**—with translucent, porcelain-like appearance.
55. **Prosogyrate**—the condition of a shell when umbones are directed anteriorly.
56. **Protobranchiate type**—gill demibranchs simple, formed of leaf-like filaments loosely connected by sparse ciliary junctions.

57. **Radial**—diverging from umbo, like the spokes of a wheel.
58. **Rostrate**—with a beak-like projection (rostrum).
59. **Sculpture**—relief pattern developed on the outer surface of the shell; sculpture is overlain by concentric growth marks corresponding to various positions of shell margins during growth.
60. **Scabrous**—rough, file-like.
61. **Scale**—localized projection on the outer surface of shell, commonly situated on a ib.
62. **Septibranchiate type**—gills absent, replaced by a muscular horizontal partition (the "septum". pierced by small pores. This structure enables a carnivorous nutrition and is encountered in a group of predominantly deep-sea bivalves (*e.g.*, Cuspidariidae).
63. **Siphons**—extensible, tube-like projections of the posterior marginal region of mantle, forming 2 openings for water inflow (inhalant siphon) and outflow (exhalant siphon).
64. **Socket**—recess of the hinge plate, for reception of a tooth of opposite valve.
65. **Tooth**—shelly projection from the hinge, received in socket of opposite valve; cardinal teeth are close to umbo, whereas lateral teeth are set apart from these, anteriorly or posteriorly.
66. **Umbo**—the first formed part of a valve, usually above the hinge.
67. **Umbonal reflection**—expansion of the internal dorsal margin which is folded over the umbones in Pholadidae and Teredinidae.
68. **Valve**—one of the main shelly halves of a bivalve.

References

Achatina fulica Krishnachandra Ghose, *Bom. Nat. Hist. Soc.*, 1963 Vol. 60 (1): 228 – 232.

Arca inaquivalvis, Bruguiere, Encyclopedie, Methodique, Vers. I, 1792.

Ariophanta laevipes, Tonapi and Leela Mulherkar, *J. Bom.Nat. Hist.* 1963, Vol. 60(1), p.116.

Astraea semicostata, Pilsbry in *Tryon, man, cinch.*, X, 1889, P.227.

Avril Bourquin, Ross Mayhew and many contributors 1999. (Man and Mollusc Uses of Shell-Bearing Molluscs - Past, Present & Future.

Bithynia stenothyroides, Tonapi and Leela Mulherkar, *J. Bom.Nat. Hist.*1963, Vol. 60(1), p.108.

Burssa spinosa, Apte, D (1998) The book of Indian shells, *Oxford University Press* 115.

Bursa margritula, Gravely, Bull. *Mad. Govt. Hus. (Nat.Hist.)* V, No 2, 1942.

Carbondale, Illinois, Land snail identification and conservation workshop. June 30 - July 3, 2008.

Cantharus undosus, Linne, *Syst. Nat. Ed.* X, 1758, P. 740, No 409.

Cardium setosum Red fern, *Ann. Lyccum. Nat. Hist.* New York, IV 1948, P 168.

Cardita bicolour, Hornell, *Common molluscs of South India*, *Mad. Fish. Bull.*, *XIV*, 1921, P. 177.

Cerastus moussonianus, Petit, 1851, Jorn, Conchyl. Vol. 2; 266.

Cerithium Citrium, Sowby, Thes. Conchyl., II, 1855, P. 179.

Cerithium obelisous, Hornell, *Common Molluscs of South India, Med. fish. Bull.*, XIV, 1921, P.110.

Cerithium scabridum Reeve, *Conch, Icon.*, XV, 1866, *Cerithium*, Pl. viii.

Cerithium variable, Landmarck, *Aniom, Sans, Vert.(Deshayes' edit.)*, 11 Vol IX, 1843, P. 302.

Collumbella vulpecula, Sowerby, *Thes. Conchv.*, I 1847, Pl – 38.

Corbicula regularis, Tonapi and Leela Mulherkar, *J. Bom.Nat.Hist.*1963, Vol. 60(1), P.119.

Crassostrea cucullata, Rao, 1969, *Indian Farming* 29 (9): 41.

Crassostrea madrasensis, Rao, 1974, Chapt. II, In the Commercial Molluscs of India, *Bull. Cent. Mar. Fish.*

Crassostrea folium, Rao, 1974, Ch-II. In the Commercial Molluscs of India, *Bull. Cent. Mar. Fish. Res.* India 25: 37.

Crassostrea cristagalli, Rao, Ch. II. In the Commercial Molluscs of India, *Bull. Cent. Mar. Fish. Res. Inst.*, 25: 36 – 37.

Cryptozona (Xestina) belangeri, Tonapi and Leela Mulherkar, *J. Bom.Nat. Hist.* 1963, Vol. 60(1), p.115.

Cypraea occellata, Grevely, *Bull. Mad. Govt. Mas. (Wat. Hist.)V*, No 2, 1942, P.40.

Cypraea caurica, Grevely, *Bull. Mad. Govt. Mus. (Nat.Hist.), V*.no2, 1942, p.41.

Cypraea onyx, Linne, *Syst. Nat. Ed. XIII*, 1791, P. 1177.

Cypraea annulus, Linne, *Syst. Nat., Ed.* X, 1758, P. 723.

Cypraea tigris Gravely, *Bull. Mad. Govt. Mus. (Nat. Hist.)*, Vol. 2 1942, P. 40.

Cyprea pallid Gray, *Bull. Med. Govt. Mus.(Nat.Hist.)*, V, 1942, P. 42.

Cyclophorus involvulus, Tonapi and Leela Mulherkar, *J. Bom.Nat. Hist.* 1963, Vol. 60(1).p.105.

Cyclophorus aurantiacus Tonapi and Leela Mulherkar, *J. Bom.Nat. Hist.* 1963, Vol. 60(1) p.106.

Cyclophorus indicus., Tonapi, G. T., 1971, *J. Bom. Nat. Hist. Soc.*, Vol. 68(1), 116.

Dalal, V. M. and G.T. Pandya, Some Land and Freshwater Molluscs of Ahmedabad. Biologica, Vol VI; 29-41.

Dentalium octangulatum, Hornel, Common Molluscs of the South India, *Med, Fish. Bull.*, XIV.

Dentalium aprinum, Linne, *Syst. Nat., Ed.* XII, 1766, P. 1263.

Desai, B. N. 1971, Dominance of Mollusca in the benthic polupation of Cochin. *J. B. N. H. S.* 68(2): 355-362.

Dholakia, A. D.(1) *"Fisheries and Aquatic Resources of India"*, 2004, Daya Publishing House, New Delhi, p. 28-31.

Dholakia, A. D. (2) *"Fisheries and Aquatic Resources of India"*, 2004, Daya Publishing House, New Delhi, p. 49-58.

Dholakia, A. D. (3) *"Fisheries and Aquatic Resources of India"*, 2004, Daya Publishing House, New Delhi, p. 126 – 135.

Dholakia, A. D.(4) *"Fisheries and Aquatic Resources of India"*, 2004, Daya Publishing House, New Delhi, p. 161-166.

Dholakia, A. D.(5) *"Fisheries and Aquatic Resources of India"*, 2004, Daya Publishing House, New Delhi, p. 314.

Digoniostoma Pulchella, Tonapi, G. T., 1971, *J. Bom. Nat. Hist. Soc,* Vol. 68(1), 117.

Donex cuneatus, Gravely, *Bull. Mad. Govt. Mas. (Nst.Hist).* Vol. 1, 1941.

Fanus ater, Tonapi and Leela Mulherkar, *J. Bom.Nat. Hist.* 1963, Vol. 60(1) p. 109.

Gafrarium dispar, Chemnitz, *Conch, Cab.,* XI, 1795, P. 230 .

Gafrarium divaricata, Chemnitz, *Conch, cab.* VI, 1782, P. 317, Pl XXX.

Gulella bicolor, Hutton, Ray, 1923, *Rec.Ind. Mus.;* Vol 15; 390.

Lamellidens marginulis, Tonapi and Leela Mulherkar, *J. Bom. Nat. Hist.* 1963, Vol. 60(1), p.118.

Litterina undulate, Gray, Zool., bookhey's voyage, 1839, P.140.

Lymnaea acuminate Tonapi and Leela Mulherkar, *J. Bom. Nat. Hist.* 1963, Vol. 60(1), p. 111.

Macrochlamys pedina, Tonapi and Leela Mulherkar, *J. Bom. Nat. Hist.* 1963, Vol. 60(1) p. 114.

Mactra turgida, Gray, *Mag. Nat Hist. Ms., I* 1837, P. 371.

Murex virgineus, Gravely, *Bull. Mad. Govt. Mus. (Nat. Hist.),* V, No.2 1942, P.49.

Murex Tríbulus Var. *Trapa,* Gravely, *Bull. Mad. Govt. Mus. (Nat. Hist.),* V 1942, P.49.

Mukharjee, A., 1969, Fauna of Rajasthan, India, *Rsc. Zool. Surv. India*: 61: 403 – 435.

Nassarius fossatus, Gould A. A. 1849.

Nass hepatica, Gravely, *Bull. Mad. Govt. Mus.(Nat.Hist.)* V No w, 1942, p.61.

Natica didya, Grevely, *Bill. Mad. Govt. Mus. (Nat. Hist.), V,* no. 2 1942, p. 37.

Natica lineate, Gravely, Bull. Mad. Govt. Mus. (Nat. Hist.) V no.2 1942.

Natica tigrina, Gravely, *Bull. Mad. Govt, Mus.(nat.Hist.),* 1942, P 37.

Nerita rumphii Reeve, *Conch. Icon.* IX, 1865, Nerita, Pl xi.

Nerita maura Rechuz, *Rev. Zool. Soc. Cuvierienne,* 1841, P. 104.

Melania (Plotia) scabra, Tonapi and Leela Mulherkar, *J. Bom.Nat. Hist.* 1963, Vol. 60(1) p. 109.

Melania (Sstriatella) tuberculata, Tonapi and Leela Mulherkar, *J. Bom.Nat. Hist.* 1963, Vol. 60(1) p. 109.

Paludomus stomatodon, Tonapi and Leela Mulherkar, *J. Bom.Nat. Hist.* 1963, Vol. 60(1) p. 109.

Parreysia corrugate Var nagpoorensis, Tonapi and Leela Mulherkar, *J. Bom.Nat. Hist.* 1963, Vol. 60(1), p.117.

Peter and Partricia Newell, "Seashells a Naturalist's and Collector's guide." Published by Phaidon Oress Limited, Oxford. 1979. P. 41 – 41.

Pecten crassicostatus, Sowerby, Thes. Cenchyl, . I 1842, P. 75.

Placenta placenta, Mossess, S.T. *J. of Natural Hist. Soc.* Vol XII.

Pila globosa, Tonapi and Leela Mulherkar, *J. Bom.Nat. Hist.* 1963, Vol. 60(1), p.107.

Planorbis (Indoplanorbis) exustus, Tonapi and Leela Mulherkar, *J. Bom.Nat. Hist.* 1963, Vol. 60(1) p. 112.

Pupoides coenopictus, Ray and Mukherjee, 1969, *Rec. Zoo.Sur. India*, Vol 61; 412.

Ray, H. C., 1961. On some non-marine molluscs from South India. Treubia, 25(3): 273 -279.

Rapana bulbosa, Gravely, *Bull. Mad. Govt. Mus. (Nat. Hist.)* Vol. 2. 1942, P.48.

Sarvaiya, R. T., 1983, "Saurashtra Molluscs, Smt. P. R. Sarvaiya, Ahmedabad.

Sulcospira hugely Var compacta, Tonapi and Leela Mulherkar, *J. Bom.Nat. Hist.* 1963, Vol. 60(1), p.108.

Tellina ala Gravely, *Bull. Mad. Govt. Mus. (Nat. Hist.)* Vol. 1, 1927, P. 106.

Tibia curata, Sowerby, G.B.I, 1842.

Trochus radiatus A. Adams, Proc. Zool. Soc. London. 1851. P. 185.

Trochus tentorium Gamelin, Syst., Nat., Ed. XIII, 1791, P. 3571, No 24.

Thais margariticol Var. *Hepteagonalia*, Gravely, *Bull. Med. Govt. Mus. (Nat. Hist.)*, No 2, 1942, P 50 and 53.

Thais bufo Gravely, *Bull. Mad. Govt. Mus. (Nat. Hist.)*, V No 2, 1942.

Venus chemnitzii, Hanley, Proc. Zool. Soc. London, 1844, P. 169.

Venus reticulate, Linne, *Syst. Nat., Ed.* XX 1758, P. 687.

Vinus papilionácea, Lamark, *Anim. Sans.Vert., V*, 1881, P. 594.

Vivipara bengalensis, Tonapi and Leela Mulherkar, *J. Bom.Nat. Hist.* 1963, Vol. 60(1) p. 106.

Vivipara Dissimilis, Tonapi and Leela Mulherkar, *J. Bom.Nat. Hist.* 1963, Vol. 60(1), p. 106.

Zootecus chin, Tonapi and Leela Mulherkar, *J. Bom.Nat. Hist.* 1963, Vol. 60(1) p. 114.

Zootecus insularis, Ray and Mukherjee, 1969, *Rec. Zool. Surv. India*, Vol. 61: 413.